MARACÁ

JOHN HEMMING AND JAMES RATTER

MARACÁ RAINFOREST ISLAND

M
MACMILLAN
LONDON

Chapters 1, 2 and 5 are by John Hemming,
Chapters 3 and 4 are by James Ratter.

First published 1993 by
MACMILLAN LONDON LIMITED
a division of Pan Macmillan Publishers Limited
Cavaye Place London SW10 9PG
and Basingstoke

Associated companies in Auckland, Delhi, Dublin, Gaborone, Hamburg, Harare, Hong Kong, Johannesburg, Kuala Lumpur, Lagos, Manzini, Melbourne, Mexico City, Nairobi, New York, Singapore and Tokyo

ISBN 0-333-55261-X

1 3 5 7 9 8 6 4 2

A CIP catalogue record for this book is available from the British Library

Typeset by Florencetype Ltd, Kewstoke, Avon
Printed in Hong Kong

CONTENTS

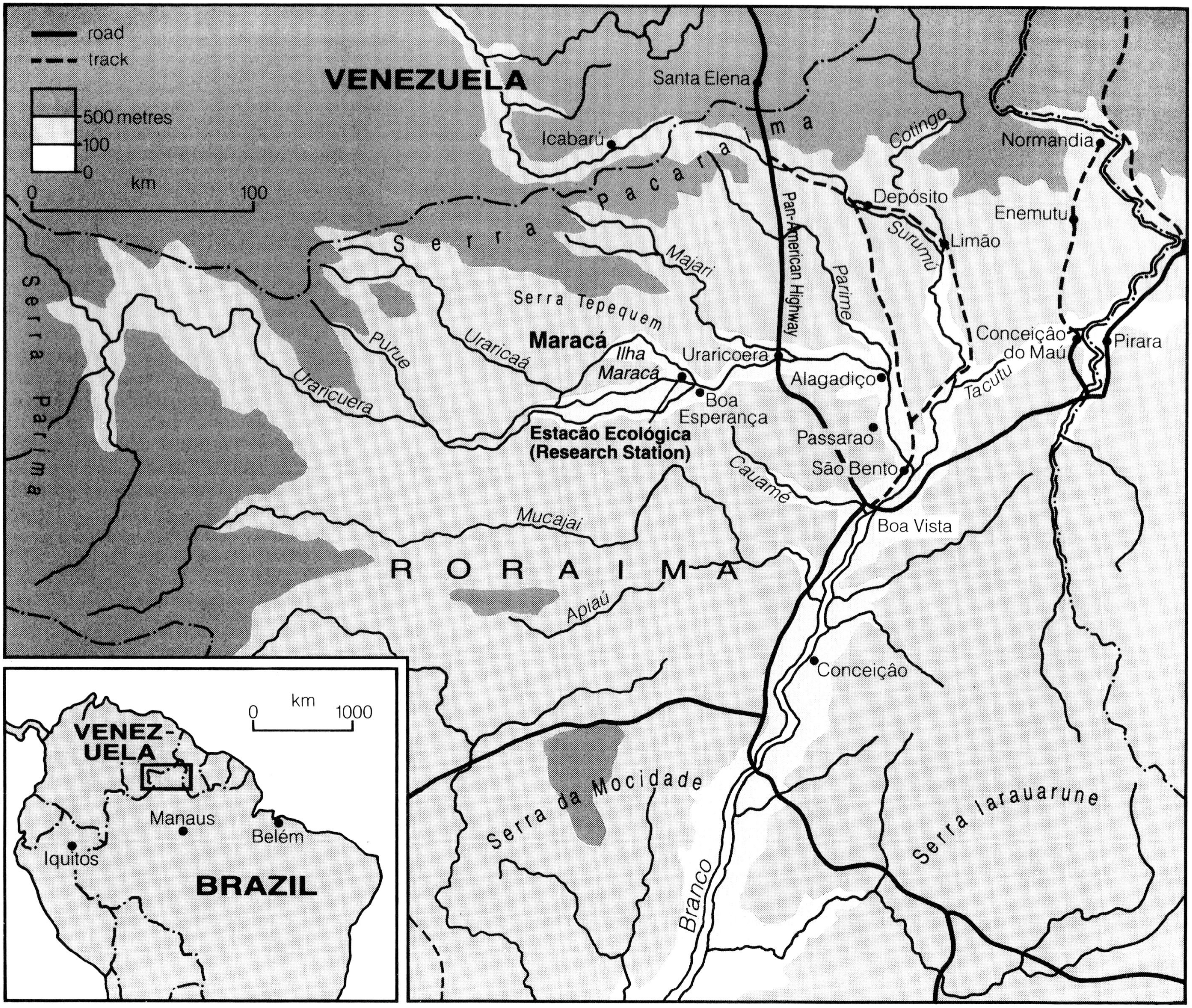
road
track
VENEZUELA
500 metres
100
0
0 km 100
Santa Elena
Icabarú
Serra Pacaraima
Cotingo
Normandia
Depósito
Enemutu
Surumú
Limão
Majari
Serra Tepequem
Pan-American Highway
Parime
Serra Parima
Purue
Uraricaá
Maracá
Ilha Maracá
Uraricoera
Alagadiço
Conceição do Maú
Pirara
Tacutu
Uraricuera
Boa Esperança
Estacão Ecológica (Research Station)
Passarao
São Bento
Cauamé
Boa Vista
Mucajai
RORAIMA
Apiaú
Conceição
Serra da Mocidade
Serra Iarauarune
Branco
VENEZ-UELA
0 km 1000
Manaus
Belém
Iquitos
BRAZIL

ACKNOWLEDGEMENTS

With such a large Project, there are a great many people to thank. Sir William Harding helped obtain the original invitation from Dr Paulo Nogueira Neto. Viscount Montgomery of Alamein and Ambassador Celso de Souza e Silva ensured that we were given permission. HRH The Prince of Wales added lustre to the Project by his royal patronage.

The steering committee chaired by Viscount Montgomery helped us to obtain funding from the donors listed in Appendix Two, and we owe all of these a great dept of gratitude. Without their generous support, the Maracá Rainforest Project could never have happened. Among so many benefactors, we should mention Dr Steven Sanderson of the Ford Foundation, John Matthews and Dino Vieira of ICI, Kenneth Light of Souza Cruz, Tim and Susie Sainsbury of the Headley Trust, Nicholas Baring, Catherine Wills of the Dulverton Trust and friends at the Garantia bank who persuaded Mercedes-Benz Brasil to lend us the '1313' truck.

Tribute is paid in this book to the dedicated efforts of the ecological station's administrator, Guttemberg Moreno de Oliveira, and his team, without whom there would have been no comfortable station for us to occupy. We also praise the administrators, led by Fiona Watson and Steve Bowles, who kept the Project running so smoothly. They were helped by a great many people in Boa Vista and elsewhere in Brazil, including the British honorary consuls in Manaus and Belém, George Clarke and Robin Burnett. The Brazilian media gave the Project good coverage and we are grateful to those who visited Maracá, particularly Ricardo Paranagua and Pedro Bial and their team from Globo Television.

Finally, we should thank the scientists, especially the splendid Brazilian researchers from INPA and elsewhere and their skilled and tireless *técnicos*. All these people *were* the Project. They worked very hard, and the Maracá Rainforest Project will be judged by the quality of their findings and publications.

We are grateful to the following for giving us permission to use their photographs for inclusion in this book:

Steve Bowles for photographs on pages 27, 29, 56, 62 and 74
Peter Edwards for photographs on page 22
Gwilym Lewis for photographs on pages 48 and 76
Arno Lise for photographs on pages 54 and 76
William Milliken for photographs on pages 25, 26, 37, 39, 47, 50, 51, 56, 62, 63, 69, 72, 73, 74, 75 and 76
Mark O'Shea for photographs on pages 28, 29, 61, 71 and 72
Mark Rickards for photograph on page 76
Duncan Scott for photographs on pages 20, 28, 29, 34, 51, and 65
Jill Thompson for photographs on pages 28, 31 and 34

Foreword
by
HRH The Prince of Wales

The combination of heat and humidity which produces tropical rainforests is found only in a belt around the equator. Although we in Britain care deeply about conservation of those wonderfully rich ecosystems and have many scientists who are expert in tropical ecology, we can study the forests and protect them only at the invitation of, and in co-operation with, countries in that region.

The Royal Geographical Society received such an invitation from the Brazilian Environmental Secretariat to work in its ecological reserve of Maracá Island in the northern state of Roraima, and I was delighted to be Patron of the resulting research project. Now that the Project's main field phase is over and its results are being published, it is gratifying that the Brazilian authorities seem to regard it as a model of international co-operation in rainforest research.

There are various reasons why the Maracá Rainforest Project was well received by its host country. It involved some thirty British scientists, but four times as many Brazilians, and researchers from the two countries found it stimulating to work together in the field. Most of the British contingent already knew Brazil well and spoke Portuguese. They were working on research desired by the Brazilians and they avoided hectoring advice about how that nation should treat the 30 per cent of the world's surviving tropical forests that it possesses. The Project's findings are being presented first in Brazil, which increases the likelihood of their being applied where it matters. At the first review conference in Manaus, all but two of the British participants delivered their papers in Portuguese, and it is hoped that all the results will be published in *Acta Amazonica*, the journal of the excellent Amazon research institute, INPA, which provided the majority of the Project's scientists.

The Maracá Rainforest Project grew to become perhaps the largest such venture ever organised in Amazonia by any European country. It is a measure of the Brazilian satisfaction with this project that it led directly to an agreement on rainforest research signed by our Minister for Overseas Development – the first such convention that Brazil has made with any country.

Apart from this political success, the Maracá Rainforest Project will be judged by the quality of its research. With so many scientists working on five main programmes and in addition a great many individual entries, the results will be extensive. Many of the findings have a direct application, while others were pure research, contributing to the general body of scientific knowledge. The Project included fine scientists from the two main sponsoring countries and, being multi-disciplinary, it provided a splendid opportunity for different types of specialist to compare notes in the field, often with fascinating results. It produced the most intensive study ever made of a forest in the northern extremities of the Amazon basin; it discovered hundreds of species new to science; and it has produced valuable information about the workings of that bewilderingly complex system, a tropical forest. One of its programmes studied settlement in a rainforest frontier region and another worked on insects which transmit tropical diseases that cause so much misery. Each of these results is valuable in itself. In sum, they add to our understanding of rainforests and, hence, to our appreciation of their importance and our ability to protect them.

CHAPTER ONE

Early Exploration of Maracá

THE first mention of Maracá was in 1740. In that year, the governor of Pará sent a large expedition up the Amazon to capture Indian slave labour. The expedition's leader, Lourenço Belforte, a planter of Irish descent, established a base on the Rio Negro and sent a powerful contingent under Captain Francisco Xavier de Andrade up the Rio Branco in search of Indian captives. Andrade later recalled that his expedition went upstream to a western tributary 'called Uraricoera which is the main trunk of the Rio Branco. Continuing our journey up this river we camped on an island from which I sent off two contingents overland along the two banks of that river in search of Wapixana and Makuxi Indians, whose villages are near hills that lie in the midst of these plains.' The northern contingent went across the savanna for a month. It was led by Captain Francisco Ferreira, a man who had been exploring the Rio Branco and trading with its Indians for twenty years or more; he had with him four chiefs from the middle Negro with their men. The southern contingent under Domingos Lopes was even more ambitious: it travelled for six weeks to the forests of the upper Catrimani and saw a turbulent headwater of the Uraricoera.

Other Portuguese expeditions went far up the Rio Branco and may well have explored the Uraricoera as far as Maracá. The official slaving troop of 1736 under Cristovão Ayres Botelho of Maranhão went up the Branco, and so did the last such troop, that of José Miguel Ayres in 1748. Meanwhile, since the early eighteenth century the Paraense Captain Francisco Ferreira had often gone up the Branco and Uraricoera and 'captured an immense quantity of heathen [Indians] for our settlements on the Rio Negro'. A royal law of 1755 abolished enslavement of Indians. An expedition of 1766 under Lieutenant José Agostinho Diniz went up the Uraricoera simply to explore and assert Portuguese sovereignty over that remote northern part of Brazil. But it 'returned without any fresh information'.

Surprisingly, the first European settlements in what is now Roraima were made by Spaniards from Venezuela, and one of their first two settlements was alongside Maracá Island. It was surprising that the Spanish should have attempted the first permanent villages, since the upper Rio Branco had been penetrated for half a century by Portuguese from Brazil to the south and by Dutch from Guiana and Surinam to the northeast. The Portuguese had a relatively easy route up the broad Rio Branco; the Dutch came across open plains from the upper Essequibo. However, the Portuguese and Dutch came in search of Indian captives and neither nation tried to settle the region.

In the mid-eighteenth century the Spaniards, who for two centuries had been established on the lower Orinoco in Venezuela, found themselves in bitter conflict with Carib-speaking tribes in the hills to the south. A series of expeditions and missions pushed Spanish rule southwards up the Caroni and its tributary the Paragua. In 1772 Indians showed two Franciscan friars how to cross the Pacaraima hills into what

Fishing at sunset.
(*Haraldo Palo*)

is now Brazil. The missionaries were surprised to find that the plains of the upper Rio Branco contained many Caribs who had fled from Spanish missions. These included the Sapará tribe who lived north of Maracá and along the Amajari River. These Indians told the friars how they regarded the three European nations:

> The Portuguese came here on the pretext of exploring, but as they took their leave they seized and carried off many people. If the Spaniards went there, they would probably do the same. But the Dutch did not do this: instead, they purchased slaves from the Indians and gave them clothing and ammunition for hunting.

The Franciscans were warned to leave by a chief who had escaped from a Spanish mission and hated his former masters. Nevertheless, the missionaries returned in 1773 and were killed. An Indian said that their deaths were 'caused by the imprudence of those missionaries who had come to penetrate those dominions'.

Indian fears were justified, for Spanish troops soon followed the missionaries. In 1773 the black Sergeant Juan Márquez de la Zapata led a force of thirteen soldiers and twelve Indians from the Paragua, through the Pacaraima hills, and on to the Uraricaá. Zapata managed to descend many rapids of the upper Uraricaá. Two days downriver, he founded a settlement called Santa Rosa after the patron saint of Lima. Four months later Zapata moved on, down to the Uraricoera River and past the northern edge of Maracá Island. He founded a second fortified settlement called San Juan Bautista on a creek at the point where the forests gave way to Roraima's natural savanna grasslands, midway between the Tipurema rapids and the eastern tip of Maracá. More contingents of Spanish troops crossed into Roraima by this route in 1774 and 1775. They pushed on from San Juan Bautista, down the Uraricoera and eastwards up the Tacutu, hoping to find the rich legendary kingdom of El Dorado in the plains towards the Rupununi.

Meanwhile, by pure chance, the Portuguese authorities on the Rio Negro came to hear about these incursions into territory that they regarded as theirs. A Belgian deserted from a Dutch fort on the Essequibo and made his way southwestwards to the Tacutu and Branco. Indians told him about the Spaniards and he spent some days with them at San Juan Bautista. He then continued his flight – Paravilhana Indians led him across the plains to Caracaraí, whence he sailed alone down the Rio Branco. The Portuguese reacted vigorously. The magistrate Francisco Xavier Ribeiro de Sampaio wrote from the Rio Negro to the Governor of Pará: 'We must consider the chance arrival of this stranger as extraordinarily fortunate. Spaniards on the Rio Branco, Sir, is a novelty as thunderous as it is dangerous!' Troops were sent up the Amazon from Belém. In October 1775 a force of fifty-eight Portuguese soldiers and over a hundred Indians paddled up the Rio Branco led by Captain Felipe Sturm, a German military engineer in the Portuguese service.

When he reached the upper Branco, Indians told Sturm that some Spaniards were on the Tacutu and others in their fort on the Uraricoera. Sturm took most of his men up towards San Juan Bautista. He sent two soldiers who were skilled woodsmen to examine the Spanish fort. The scouts were well received by its small garrison and reported back that it was a log building with inner and outer walls having four or five cannon that fired stone balls mounted at embrasures. Nearby were two new mud-walled dormitory buildings, a manioc plantation, and an orchard. Captain Sturm arrived next day, 14 November 1775. He refused to parley with the Spanish commander, Sergeant Juan Antonio Cuello. Instead, he arrested the thirteen men of the garrison and sent them down to the Rio Negro in irons. The settlement was destroyed and its cannon removed. Sturm then started building a fort called São Joaquim at the junction of the Tacutu, Uraricoera and Branco Rivers. During the following weeks, he captured a further twenty-six Spaniards – by deception, without firing a shot – and all were sent downriver as prisoners.

Sturm's swift campaign, and particularly the building of Fort São Joaquim, confirmed Brazilian control of the upper Rio Branco, modern Roraima, for evermore. That sovereignty was recognised in the Treaty of San Ildefonso of 1777, which confirmed the earlier Treaty of Madrid of 1750. Both treaties declared that the watershed between the Orinoco and Amazon river-basins was the frontier between Spanish and Portuguese possessions in this part of South America. Since they fixed the frontier on a geographical feature, these far-sighted treaties have stood the test of time. Brazil's northern border has never been disputed for over two centuries.

Immediately after the building of Fort São Joaquim, the Portuguese authorities tried to settle Roraima's Indians in new villages on the banks of the main rivers. The Indians were persuaded to move into these *aldeias* by presents of metal tools – which the Indians valued greatly – and beads and clothing. There were soon six villages; the largest of these was Nossa Senhora da Conceição on the south bank of the Uraricoera, not far below Maracá Island. Conceição contained 372 Indians, mostly Carib-speaking Paravilhana and Irimissana Sapará. But the attempt to colonise the native tribes soon went wrong. The presents that had enticed the Indians into the villages ceased. Soldiers were installed as village directors and they started to demand labour from the Indian men and women. Their charges were expected to provide quotas to paddle canoes or to help build the fort; others were sent to fish or collect turtle eggs to enrich the director. The Indians naturally resented this forced labour, which was wholly alien to their own communal society.

Manoel da Gama Lobo d'Almada, a Portuguese officer and future governor of Rio Negro, realised that

> to remove these uncontacted tribes from the forest, where in their manner they live more comfortably than among us, we must persuade them of the advantages of our friendship. We must feed and clothe them, not fatigue them by demanding more work than they can do, and make sure they are paid promptly and without usury what they are promised and owed, which they have earned with the sweat of their brows and sometimes at risk of their lives.

This wise advice was not followed. Instead, there were clumsy attempts to change the Indians' way of life. A commander of the fort wrote that 'we try, gently enough, to forbid the abominable customs that they have always had – such as burning the bodies of the dead in the houses in which they died, or allowing each man to have as many wives as he pleases. They are also very astonished to be forbidden to anoint themselves with *urucum* [red anatto] body paint and to be forbidden many other perverse and abusive customs that they are very loath to abandon.'

The village Indians started to refuse to work for their directors. The fort's commander tried to arrest chiefs from a number of villages but, as a Portuguese explorer wrote, 'the hatred among the Indians of Conceição [on the Uraricoera] grew steadily. They saw the multiple arrests of their chiefs added to their other grievances – the frequent labour quotas imposed on them and the violent abuse they suffered from the soldiers – and almost all decided to withdraw; which they did.' In 1781 most of the Indians of the Rio Branco villages fled, killing some soldiers and often burning the settlements. The government in Lisbon decided on a conciliatory response, and a royal pardon for the Rio Branco rebels was proclaimed in 1784. Over a thousand Indians were persuaded to enter a new set of six villages on the banks of the Rio Branco, but these villages were farther downstream and there was no attempt to revive Conceição on the Uraricoera. The second experiment lasted no longer than the first. The same errors were repeated and in 1790 there was a second 'rebellion' which started when Makuxi working in a fish factory near Fort São Joaquim went on strike and threatened their overseer with knives. This time the authorities reacted harshly. The inhabitants of all but one of the villages were banished to new locations hundreds of kilometres away, on the Madeira, Solimões, and Amazon Rivers. All the Paravilhana and most of the Sapará tribes were exiled in this way. A third revolt by Roraima's Indians ended in a pitched battle and massacre in 1798, and there was no further attempt to colonise them in villages.

The treaties of 1750 and 1777 roughly gave Brazil its modern boundaries, an immense area covering almost half the South American continent. The Portuguese now set about exploring and mapping the boundaries of this vast empire. In the 1780s they sent teams of boundary commissioners and surveyors up many of the remotest rivers in Amazonia, including the tributaries and headwaters of the Rio Branco. Some of Brazil's greatest explorers took part in these expeditions, guided and paddled by local Indians. It was a golden age of exploration, and three expeditions in quick succession went up the Uraricoera River past Maracá Island.

In March 1781 two brilliant surveyors, the army engineer Major Ricardo Franco de Almeida Serra and the cartographer Dr Antonio Pires Pontes, made this journey. They passed the mouth of the Amajari and the populous village of Conceição. On 18 March they passed the Maracá Channel 'which is wide, so that it looks like the main stream of the Uraricoera River'. They then visited the remains of the Spaniards' fort, San Juan Bautista. Since their journey was near the end of the dry season, they had a difficult time passing the many rapids on the Santa Rosa Channel north of Maracá Island. On 20 March they spent all day passing the Tipurema rapid. They struggled all the next day to overcome two more rapids, advancing only a few kilometres. On the 22nd they passed the mouth of the Idume (Trairão) stream and camped below the Fumaça rapid. It took them the following morning to ascend this and the next four days to conquer more rapids and races. Almeida Serra wrote that they spent the 28th and 29th of March passing three more rapids 'with much danger. My canoe sank three times and on one occasion I was left hanging from a tree.' The expedition tackled fourteen big rapids in six leagues (36 kilometres) before finally reaching the mouth of the Uraricaá River. The larger Uraricoera was blocked by 'horrific rapids, so frightening that I judge navigation up such a river to be extremely difficult. We therefore returned and continued our journey up the Uraricaá River.' The explorers had seen the magnificent Purumame waterfall at the western tip of Maracá. Five days' journey up the Uraricaá, Almeida Serra and Pires Pontes reached the abandoned Spanish settlement Santa Rosa, and they turned back a short distance upriver of it. Returning past Maracá Island, they endured 'a thousand hardships in the rapids, in one of which all the canoes swamped and were recovered two kilometres downstream without benches or awnings'. They decided to investigate the Maracá Channel, going up it for three leagues (18 kilometres) on 14 and 15 April.

> This river is very broad and full of water. It runs between extensive plains. After going up it for this distance, during which two streams entered from the right, we came upon a multitude of large and small islands, after which there was a sizeable rapid. We passed this and navigated between other larger islands which make the river very broad, as far as another waterfall. We did not pass this, both because we had no guide and because time was too short for this exploration which is not of great importance.

These two explorers also investigated the Amajari, Parima, Tacutu, and Maú Rivers and drew a fine map of the region.

Five years later, in July 1786, the naturalist Alexandre Rodrigues Ferreira also went up the Uraricoera. Rodrigues Ferreira was the first Portuguese scientist to work in Amazonia. A botanist and keen ethnologist, he left vivid descriptions of his journeys, reporting on many aspects of Amazon Indians and wildlife. He moved up the Tacutu and Surumu Rivers towards Mount Roraima and then turned westwards. He went up the Uraricoera only as far as the site of the Spaniards' short-lived settlement San Juan Bautista on the north bank of the Santa Rosa Channel. It had been located between the Caya-Caya and another stream, at a place where the main river narrowed and was broken by rocks. Rodrigues Ferreira also mentioned that the Maracá Channel had been explored by the black Spanish sergeant Juan Márquez de la Zapata, who had then marched overland for ten days to the upper Mucajaí in search of Indian captives. But the fierce Wayumara Indians who lived near the great rapid on the Mucajaí (now the site of the Paredão dam) killed Zapata and his Spaniards, probably in 1775. A decade later, the Portuguese soldier Duarte Joseph Migueiry – a famous woodsman and explorer – also went up the Maracá Channel for three days, past innumerable rapids.

A year after Alexandre Rodrigues Ferreira's visit, the great explorer Manoel da Gama Lobo d'Almada also went up the Uraricoera. He described the river above the Spanish fort as full of 'laborious and dangerous rapids; and above the confluence of the Uraricaá it is very difficult to navigate, for it immediately encounters the hills from which it receives its waters, and climbs rapids after rapids that make a passage quite impossible'. Gama Lobo d'Almada went right up the Uraricaá as far as the point where

> it gave me passage, always between and over hills, climbing continuous rapids, until I disembarked from the Araicuque stream, which enters the Uraricaá from the west, and proceeded overland. Making my way through the hills of the *cordillera* [Pacaraima mountain range] in a northwesterly direction, I reached the headwaters of the Uraricaá, and from there I advanced and crossed to the north slope of the *cordillera*, where the waters flow towards the Orinoco.

There, on 12 April 1787, on the Anocapre headwater of the Paragua, disaster struck.

> Heavy rain started at two o'clock at night. By five o'clock, it brought such a great torrent of water down from the mountain that in a few minutes the valley was flooded. Those of us who saved ourselves most promptly emerged with water up to our knees – because of which accident we called that place the Valley of the Flood. We climbed the flank of the mountain. The darkness of the night increased our confusion and I wished that day would break so that we could search for some of our missing belongings. I was most concerned about the sextant that the mathematical doctor Euzebio Antonio de Ribeiros had lost, for this was before he had taken the latitude of the place, which was what he and I most regretted. The torrent had carried off that instrument; but happily it lodged in a tree where we found it! The sextant still worked and, although the surveyor lost a compass and watch, he fortunately had spares.

Gama Lobo d'Almada thought that he had survived relatively unscathed. But when he crossed back through the hills, he learned that the same rains had caused a flash flood on the southern side. Two of the expedition's canoes had sunk, one of them containing the orders from Governor João Pereira Caldas. There was worse to come near Maracá Island. Ten days later on 22 April:

> when we were descending one of the largest rapids on the Uraricoera, I was shipwrecked with my canoe (the same in which I had left Barcellos). This rapid has seven big drops besides many other smaller ones. I was lost in the first of them, so that the canoe went on smashing into the other falls and disintegrated. With it I lost the maps that Your Excellency [Governor Caldas] had given me . . . Your Excellency has probably heard that I cannot swim. The soldier Manoel de Sousa Monteiro started to rescue me, but he was almost drowned. The brave soldier Alberto Serrão de Castro without hesitation jumped into the midst of that raging water: he luckily caught me and disentangled me from [Sousa Monteiro]. He seized me by a wrist and held on to the wrecked canoe with his other hand. He continued to support me as the current swept us down, until I was saved on a rock in the middle of the river, far below the rapids. A small canoe paddled by soldiers and an Indian came to get me there . . . Those who watched the shipwreck say that I struggled in the water for over half an hour and that on four occasions I disappeared but always came up again with Serrão. I received a heavy blow on one leg when I hit a rock; it is healing well, although to this day I have to bleed it. When the canoe hit the rocks, I was badly bruised on my right arm and it is all black; and my left leg is very badly wounded, although not dislocated. The cuts are big but not very deep. The swelling is going from my arm and my leg is less inflamed, and I am in far less pain than during the first three days when I could hardly stand it.

Gama Lobo d'Almada treated his wounds by blood-letting, plunging them in wine, binding them, and covering them in unguents while he himself drank brandy. He asked the Governor for some clothing and a hammock, 'a few insignificant things, but necessary for someone who was shipwrecked with all he possessed'.

He went on to perform many more services and became one of Rio Negro's best governors. Gama Lobo d'Almada is also credited with having introduced the first cattle into Roraima; he extolled the great potential of that northernmost part of Brazil.

The splendid burst of exploration during the 1780s was followed by half a century of stagnation. All attempts to settle Roraima's Indians or to explore and map its rivers were abandoned. There were very few people in the region apart from the small garrison of Fort São Joaquim and a handful of cattle ranchers. Even without husbandry, the cattle multiplied: the few animals introduced in the 1790s increased to three thousand penned and perhaps five thousand wild by 1840.

In September 1838, 34-year-old explorer and naturalist Robert Schomburgk reported to Henry Light, Governor of British Guiana, 'conformable to the instructions received from the Royal Geographical Society, when entrusted with the command of an expedition in the interior of British Guiana'. The Council of the RGS had decided in 1834 to send an expedition to explore and map the southern part of this colony, which Britain had acquired from the Dutch after the Napoleonic wars. As leader of this venture, the RGS appointed Schomburgk, a German who had impressed the Society by the quality of a map he submitted of a Caribbean island. In 1835 he explored the Rupununi River in southern Guiana and in 1838 he was the first

European to reach the source of the Essequibo. He now explained that he was about to embark on an expedition whose 'object is, according to the direction of the Royal Geographical Society, the attempt of discovering the sources of the River Orinoco, which are still a geographical problem'. Schomburgk spent the rainy season of 1838 at Fort São Joaquim, and he wrote, 'I cannot speak too highly of the civilities which we received from the Brazilian authorities'. Although Amazonia was then in the throes of the bloody Cabanagem Rebellion, the commander of the Upper Amazon instructed his brother Pedro Ayres to help the British explorer in every way. Schomburgk wrote that 'on my application to Senhor Ayres I have been provided with six Indians and a soldier from the garrison, and with this reinforcement added to my own party I consider myself strong enough to commence our journey'.

Schomburgk set out in October 1838. He moved northwards up the Maú and Cotingo towards Mount Roraima, climbing higher up the mountain than any previous explorer. After a month investigating the surroundings of Roraima, Schomburgk struck southwestwards across the savanna. On 5 December he reached the Paparu (now called Pau-Baru) stream, which flows into the Uraricoera near the northern tip of Maracá. Purokoto Indians helped Schomburgk assemble four small canoes, and a Maiongong explained to him the geography of the Orinoco, Mavaca, and the Casiquiare canal linking the Orinoco to the Negro. Schomburgk admitted that 'from this information . . . I thought it advisable to follow the Paraima [Uraricoera] instead of the course of the Uraricapara [Uraricaá], as I had first intended'.

When Schomburgk entered the Uraricoera, he immediately found that his frail canoe took on water in the rapids. At the mouth of the Trairão he had his men cut down a *Cecropia peltata* tree 'out of which two side-planks were made and shortly added to our washboards, which, though it did not accelerate our way, at any rate rendered our passage more safe, and saved the people from getting wet'. This seasoned explorer was laconic about the hardships of passing the northwestern edge of Maracá. He wrote, 'A succession of falls made our progress very slow: at [four] of them we were obliged to unload the corials [canoes] and carry the luggage overland.' He remarked that in three days he advanced only fourteen miles but gave no details of the horrors of the rapids. His decision to push on up the Uraricoera – the first European to attempt this difficult route – meant climbing past the formidable Purumame waterfall.

> The Purumame Imeru is certainly one of the largest falls in Guayana, vying in size and magnificence with William IV's cataract on the Essequibo, and the falls of the Corentyn . . . Narrowing to about 50 yards, the river divides into two streams and precipitates itself from a height of 40 to 45 feet: the whole width of the river at the fall was not more than 10 yards when we passed it . . . A little beyond occurs a second fall of about 25 feet, making altogether a descent or fall in the river of from 70 to 75 feet from these cataracts. We had no alternative but to carry our corials over the range, which was about 350 feet above the river.

The Indians' portage path was extremely steep, but Schomburgk's men hauled his boats and equipment up and over the hill in half a day. On a brown rock outcrop at the foot of the trail, the explorer left an elegant 'S' carved in a circle. On the following day, Schomburgk passed the mouth of a branch of the river called Maracá – the Makuxi for 'rattle' – and his Indians explained that this channel rejoined the Uraricoera some sixty kilometres downstream, to enclose the island of Maracá.

A short distance above the western end of Maracá Island, Schomburgk's party visited a village of some forty Sapará Indians. 'The men of the tribe were hideously ugly . . . They seemed to suffer, most of them, from inflammation in the eyes, many of them squinted horribly, and others were evidently dropsical. Their voices were squeaking and very disagreeable . . . The women were a striking contrast, and really very pretty; one girl I thought the prettiest Indian I had as yet seen.' These Sapará lived in a neat circular *maloca* (communal hut). After five more days 'toiling with much labour and fatigue up a succession of falls and rapids', the explorer visited a village of Wayumara, the tribe that had killed the Spaniards on the Mucajaí. 'The captain received us, sitting on a low stool, surrounded by his men, all armed with war-clubs.' But the tribe was friendly and traded manioc after giving Schomburgk a cigar to smoke as a peace offering. The expedition toiled westwards throughout the month of December 1838, through the territory of the Krixana Yanomami and

Freshwater biologists worked in savanna wetlands. (*Dudu Tresca*)

then left the Uraricoera to push up the Araracá. The canoes were eventually abandoned, and for a week at the beginning of 1839 the men walked across the Parima hills to the Marevari River in Venezuela. Schomburgk duly made his way across to the Orinoco, through the Casiquiare canal, down the River Negro, and up the Branco. He returned to Fort São Joaquim on 22 April, having completed the remarkable circuit of some 3500 kilometres in seven months and two days – precisely the time he had predicted he would take.

Soon after Schomburgk's epic expedition, there was some missionary activity on the lower Uraricoera, not far downstream of Maracá. In 1839 a Carmelite friar, José dos Santos Inocentes, was sent to establish a mission on the upper Rio Branco. He started his mission near the Tacutu, but in 1841 he moved it westwards to the Barico hills overlooking the Uraricoera, close to the long-abandoned mission of 1776–81, Conceição. For a time, the friar had acculturated Makuxi and Wapixana living in this mission, which was called Porto Alegre, but he was too old and ill to make further journeys to contact more Indians. In 1846 Friar José was moved to the upper Rio Negro. His successor 'was not endowed with the vocation necessary for directing an establishment of that kind. He immediately displeased the Indians, who gradually withdrew themselves.' When another missionary came in 1849, he found the 119 Indians of Porto Alegre living wretchedly, and three years later, the provincial government closed the mission definitively. There was no further missionary effort in Roraima during the nineteenth century, and visits by parish priests from the Rio Negro were very rare.

It was forty-four years after Schomburgk's journey before explorers again investigated the Uraricoera. In August 1882 the government of Brazil sent a team up the river to survey the Brazilian frontier with Venezuela. The team was led by Major Dionísio Evangelista de Castro Cerqueira and included three scientists, Drs Antônio de Souza Dantas, Feliciano Antônio Benjamin, and José Jardim. It had nine canoes and twenty-five Indians of the Wapixana, Makuxi, Sapará, and Wayumara tribes; farther upriver it was also helped by Purukotó guides. On 28 August 1882, the expedition passed the tip of Maracá, which was now known to be an island 'larger than many of the sovereign states of Europe'. On the Santa Rosa Channel, the explorers found the remains of the short-lived Spanish settlement San Juan Bautista, now marked only by 'an old cannon, half buried in the interior of the forest, with a few cannon balls already consumed by over a century of prolonged rusting'. Major Castro Cerqueira noted that

> at the western extremity of Maracá Island begins the wilderness that stretches onwards into the unknown, as far as the impenetrable regions of the Parima hills and upper Orinoco. Civilised Indians never dare remain in those solitudes, whose solemn silence is broken only by the war cry of their savage enemies. At Caya-Caya [on the Santa Rosa Channel] the open plains cease and luxuriant forest, stupendous and majestic, begins. The land changes its appearance and the river also assumes a different aspect. Sluggish, ill-defined and almost devoid of islands [below here], it now becomes full of rapids, turbulent, foaming, confined between ranges of hills and lofty banks, its bed strewn with innumerable blackened boulders and covered with islands crowned in a profusion of vegetation under whose benign shade that precious plant vanilla and a great variety of orchids spring up in remarkable abundance.

This expedition battled up the rapids alongside Maracá Island and then went up the Uraricaá River. At the end of September it crossed the Pacaraima hills and entered Venezuela with the help of Waika (Yanomami) Indians. Major Castro Cerqueira had investigated the great Purumame falls that blocked progress up the Uraricoera. He was appalled by what he saw. He wrote of the upper Uraricoera:

> That remote, mysterious, and uninhabited wilderness, bereft of resources and dominated by wild hordes of Maracana, Krixana, and many other tribes that infest the solitudes of the Parima hills, will remain inaccessible to civilised man, shrouded in the mystery in which it has existed until this day. Any rash man who risks penetrating such inhospitable solitudes with a regular expedition will either pay for his temerity with his life, or will return without having reached his destination.

The major did not realise that Robert Schomburgk had already done this, almost half a century previously.

Towards the end of the nineteenth century, there was an increase in cattle ranching in Roraima. Although by 1885 there were still only twenty thousand head of cattle on thirty-two private ranches, settlers were establishing themselves all along the savannas on either bank of the lower Uraricoera. One such was Manuel Galvão from Rio Grande do Norte, who came in about 1880, settled in a house called Livramento Maracá, married a Wapixana woman and had a large family. Galvão said that when he arrived, there were many Indians of different tribes there and on Maracá Island. But a villainous rancher called Hollanda Bessa then started a *fazenda* named Santa Rosa on the north bank of the Uraricoera near the eastern tip of Maracá Island.

> Bessa chased the Indians from their homes and plantations, claiming the land belonged to him. He burned their huts in the midst of the rainy season. The poor people fled from him into the forest, but fell ill from fevers and with no protection against the wet, many died, especially children. As a result, only piteous remnants of the tribes are now [1910] to be found around Maracá. Bessa was charged, but released on the testimony of some good friends he had in the ruling party. The Indians are too indolent to kill him.

In 1910 a 38-year-old German anthropologist, Theodor Koch-Grünberg, decided to try to repeat Schomburgk's extraordinary exploit of going up the Uraricoera River and crossing into the upper Orinoco basin. Koch-Grünberg was a brilliant anthropologist who had previously worked among the Indians of the upper Xingú and on the Uaupés tributary of the upper Rio Negro. He reached the eastern tip of Maracá on 3 December 1910. He landed on the northern shore of the island and walked inland with a German friend to visit a couple of Indian families.

> We went for a good distance through light woods, with which the majority of the eastern end of the island is covered, then over a small savanna to a roomy, cool hut in good condition, which was inhabited by a good many hens. You could see that orderly people lived there. [The Makuxi] Felipe hurried out to greet us and serve us delicious maize *cashiri*. [His friend the Wayumara] Manduca was off hunting: he soon returned with two fowls. One unpleasant aspect of this place was countless *piums* [small biting flies], which we encountered at the beginning of the rapids. They are said to be even worse upriver.

There were very few Indians left around Maracá at the time of Koch-Grünberg's visit. Most had died – from disease, from ill-treatment by the whites, or in intertribal fighting. The Sapará tribe, which had three hundred members in Schomburgk's day, was reduced to one small *maloca* on the Maracá Channel. The Wayumara were reduced to the man met by Koch-Grünberg and his brother, who lived on the Tacumá stream on the Santa Rosa Channel.

> In the past there was a big Indian village here called Santa Rosa, a former mission under Chief Ignácio. You can still see its extensive ruins visible on the banks. The villain Bessa removed it, so that it is now difficult to get even a couple of men for our onward journey.

Farther up the Santa Rosa Channel, Koch-Grünberg was shown the site of a former Sapará plantation, marked by the clear trunk of an *ambaúva* (*Cecropia*) tree. This was the work of the Sapará whose village Schomburgk visited in 1838. 'When the Sapará all died out, the Tauripang took their place and built a *maloca* near the riverbank; but this also disappeared long ago.'

Theodor Koch-Grünberg's expedition set off in three canoes and soon passed the hut of an old black called José Victôrio, 'the last representative of civilisation'. For all the settlers, the known world ended at Maracá. All they could tell the German explorer were the names of the two channels. They avoided the Maracá Channel 'for fear of the wild Marakana Indians who make its banks unsafe' and knew nothing of the Uraricaá River. All were convinced that Koch-Grünberg would never reach the Orinoco: he was bound to die in a rapid or from an Indian arrow.

At midday on 9 December the explorers reached the first serious obstacle: the Tipurema rapids near the northern tip of Maracá Island. They were travelling in the midst of the dry season, and the Indians told them that the river was unusually low, so they had a terrible time. 'The river rages over a barrier of huge boulders', which forced them to unload and portage. It took them four hours to haul the empty canoes over the rocks with cables. They camped on an island in the midst of surging waters opposite the mouth of the Paparu (Pau-Baru) stream and spent the following morning ascending the Paparu rapids, using cables and poles but without having to unload. They passed the mouth of the Trairão and then started to confront all the fierce rapids along the northwest side of the island. Koch-Grünberg complained that

> it is endlessly tiring to drag the heavily laden boats over the rocks and between the boulders. Rocks crowd in on us everywhere, often leaving only the smallest passage clear between them. The flood waters raced through a tangle of little islands and poured out in waterfalls.

It seemed a wonder that the boats survived such punishment. After much hauling with ropes, the expedition reached a massive rapid that the Indians called Monarupá. Waves struck at the loaded canoes. The men struggled forward until they were close to the roaring fall, but they could not get the boats through it. 'We have to return and go through a more southerly channel, but this is also terrible. A sea of rocks! With massive exertions and wild shouting, the boats were finally brought up it.' At midday the team rested on a rocky island and the German anthropologist had a bowl of Maggi soup – a welcome change from fish. After lunch there was:

> a continuous tangle of rocks: innumerable little islands to right and left, innumerable channels through which the water pours thunderously. The work is unspeakably hard and we advance only very slowly. It continues like this as far as the Arukaima rapids, which start as a high fall and emerge in an endless confusion of boulders, islets, occasional small waterfalls, rushing rapids, and spluttering whirlpools. The river is very wide here and interspersed with high boulders of mica schist which rise darkly out of the white spray of the billows. It is a wildly romantic picture. We work laboriously with paddle and ropes through the surging waves as far as a rocky island, where we camp. You can make yourself heard only by shouting. The roar of the waterfall drowns all noise.

Next day, 13 December, was spent portaging loads around the rapids and dragging the boats over rocks covered in slippery weeds – and then repeating this labour at another rapid a kilometre upstream. The following morning:

> our crazy journey continues again at about seven o'clock. The river is continuously wide and level, but full of boulders and islands. The Indians say that the water level is unusually low: it would have been better in other years. Descents of water follow one another almost without interruption. Even level rapids like Uemelu occupy us for hours. The men have to roll aside large rocks to open a narrow passage for the boats.

So it continued, day after day. Finally, a trio of wild rapids were surmounted in quick succession and the expedition entered calm water. The men could scarcely believe it:

> The roaring continues in our ears: we are not used to quiet. The boats look terrible: leaks have sprung everywhere. The thwarts, which served as hand-holds when we were pushing them over boulders, have been torn out in places. We can scarcely bail out all the water that leaks in.

A consolation for Koch-Grünberg during his difficult journey was the abundance of fish – *pacú*, *curimatá*, and big *pirandirá*. His Indians also shot game birds, and he described their curious calls. The *mutum* (curassow) sounded like the growl of a dangerous beast of prey; the droll *jacami* had a repertoire of noises, from a trumpet note to the mimicry of the deep sound of some mammal; the small *socó* heron sounded like the wail of a ship's siren; the *cujubim* was like two tall trees rubbing together. The men also ate the eggs of chameleon lizards, of *jabutim* (land tortoises), and of river turtles.

In order to follow Schomburgk's route up the Uraricoera, Koch-Grünberg had to climb the formidable Purumame waterfall. He had sent Indians up the Uraricaá to fetch some Xiriana to help him over this barrier. He was duly impressed by the mighty falls, which thundered down a drop of 30 metres in two great cascades. He marvelled at the foaming white water between walls of black mica, all shrouded in mist-like spray. Maiongong Indians were said to put red pepper in the eyes of anyone who saw the Purumame falls for the first time, so that its resident demon could not attack them with fever.

Finally, the Xiriana arrived: twenty men, women, and children who had never had contact with whites and who had all the usual attributes of the closely related Yanomami. The portage path alongside the falls was well worn by constant use, but it was almost vertical. Koch-Grünberg had his men widen the trail in order to be able to take his boats. Fifteen muscular Xiriana made light work of the portage. They heaved his boats and heavy loads up the steep hill, along the lengthy trail, and past the next set of rapids. It took them only a few hours to do work that would have taken Koch-Grünberg's six other Indians two days. He marvelled: 'It is a mystery to me how those lads climbed it with such heavy loads in so short a time.'

The struggle with rapids began again immediately above the Purumame falls.

> Another labyrinth of islands and gigantic rocks opens up before us. Boiling water shoots down between them. All the boats are safely pulled up the worst part of the rapids, and I pause for a few minutes' rest, sitting on a high rock. I intend to follow overland, when there is a shout. A pair of boxes, baskets, paddles, and arrows float past me. Akuli's boat has sunk once more. He and José swim after the belongings and manage to rescue most of them.

Finally, on 29 December, the expedition passed the western tip of Maracá Island. Even the Indians were fed up 'with the constant battle against this uninhabited river, shut in by gloomy high forests, with its fearful rapids and waterfalls and its tangle of rocks between which the water disappears'. But Koch-Grünberg did battle on. He ascended the Uraricoera and crossed the hills to the Meruari and Ventuari tributaries of the Orinoco. After spending many months among the tribes of Venezuela, he descended the Rio Negro in 1913, thus completing most of the circuit achieved by Robert Schomburgk seventy-five years before.

Eleven years later, in 1924, Theodor Koch-Grünberg returned to the upper Rio Branco, this time as a member of an expedition led by an energetic American explorer, Dr Alexander Hamilton-Rice. The German anthropologist was profoundly distressed by the changes that he saw among his beloved Indians. He wrote in his journal:

> The Indians of Rio Branco are close to their end. Those who escaped the influenza epidemic that killed entire *malocas* [villages] are now being liquidated by *balata* rubber gatherers, gold prospectors, or diamond seekers . . . The few surviving Indians have been deprived of their rights and reduced to slavery. Their ingenuous happiness has gone, their solemn dances have ceased, the joyful play of children in the sandy centre of the village by moonlight has ended. Happy are those who died in time.

Shortly after this sad comment, Koch-Grünberg himself was gone – struck down by malaria and buried at Vista Alegre near the main rapids on the Rio Branco.

Despite the death of his guide, Hamilton-Rice's expedition pushed on, determined to ascend the Uraricoera to its source. Dr Hamilton-Rice was a dynamic explorer who had previously investigated the rivers of the upper Rio Negro. He brought with him two innovations: a radio, with which his expedition maintained contact with its base, and the first aeroplane ever to fly in Roraima. This hydroplane was a Curtis Sea-Gull, a lovely wooden biplane fitted with floats. It took the first aerial photographs of Boa Vista and of the forests and rapids of Maracá Island.

Hamilton-Rice's expedition reached Maracá at the end of December 1924. It enlisted a guide named Cyro Dantas. Theodor Koch-Grünberg had met Dantas in 1910 and described him as a likeable and idealistic young man from a leading family of Paraíba. At that time Cyro Dantas lived in a hut called Independência far up the Maracá Channel, remote from civilisation and surrounded by Indians. Fourteen years later, the young romantic had changed into a river trader plying the Uraricoera and married to the daughter of the brutal Hollanda Bessa. But he did know the river and its tribes. In 1922 Dantas paddled far up the Uraricoera and made friends with the Maiongong of a village opposite the mouth of the Araraca. Hamilton-Rice hired Cyro Dantas 'as *capataz* [overseer], whose mission it was to ride over the country and recruit Indians as crews for the canoes. His efforts in two weeks' time netted some thirty Indians – Makuxis, Wapixanas, Arecunas, and Taurepangs.'

Part of Hamilton-Rice's expedition camped at the Tipurema rapids near the northern tip of Maracá. The hydroplane made reconnaissance flights far up the river, taking beautifully clear pictures of the island. It had a slight accident when taking off from Tipurema and two American engineers camped on a nearby island repaired it. 'The Tipurema camp was placed in a very picturesque situation, just over the *cachoeira* [waterfall] in the midst of high trees, and a deer shot just a short time before our arrival was hanging up, being dressed.' From 13 January 1925 onwards, the expedition battled up the rapids that had caused such trouble to all previous explorers. 'The river varies from amphora-like swellings to bottleneck constrictions, and when actual rapids do not exist, there are difficult *corredeiras* [strong currents] that run like mill-races.' Much of the time paddling and poling availed nothing. The only progression possible was made by long cables carried to far points, one group hauling and pulling in on the lines as another pushed and dragged a heavy canoe over the slippery or sharp rocks that offer only precarious footholds, and which are dangerous to limbs by reason of the holes, cracks, and furrows that corrugate or break the surface. It was often necessary to unload the expedition's mass of cargo and carry it overland while the empty boats were heaved up rapids. At the Pelalaitapang rapids a punt

> was speedily swamped and wedged between great boulders in deep rapid water that necessitated two hours' difficult work with axes and saws on the part of eight men. At the next rapids, it took four hours to pull the big *batelão* [large canoe], two [smaller] canoes, and the punt past the rocks.

So it continued, with day after day of rapids, reefs, and rocks. On 19 January the expedition finally reached the mouth of the Uraricaá and camped beyond it at the foot of the mighty Purumame waterfall. Hamilton-Rice described this as:

> the largest fall in Brazilian Guayana, an impressively wild and beautiful cataract . . . The Uraricoera hurls itself over the falls to the beaches below; the whole hemmed in by a great gorge that accentuates the roar of the waters, making an impressive and inspiring spectacle, stupendous in mass and volume.

The expedition's geologist noted that the fall's 29-metre (95 feet) drop consists of three steps formed by faulting. The upper step is flanked by schists and composed of igneous rock and fine argillaceous sandstone, the middle step is quartz diorite, and the base of the fall is highly metamorphosed granitic rock with a finely grained biotite schist.

The expedition camped for a fortnight on the sandy clearing on the north bank below Purumame. It used the Indian trail to transport petrol cans and other cargo up and over the steep hill beside the falls. But for the canoes, 'a road had been broken through and constructed around the foot of the hill and carried to a short distance above the upper falls'. The heavy canoes were moved up this road on log rollers. The boats had been much damaged in the rapids and had to be repaired, but 'fortunately the neighbourhood abounded with the *sorba* tree (*Couma utilis*) whose white sap when boiled turns to a black pitch-like mass that makes excellent tar with which to finish the caulking'. The expedition marked its camp by carving an inscription and a drawing of its biplane on an outcrop of black rock. Its carving is still there, close to Schomburgk's 'S' and an inscription made two years later by General Cândido Rondon, the great founder of Brazil's Indian Protection Service.

Farther up the Uraricoera, Hamilton-Rice's men had many adventures with rapids and encounters with friendly Indians. They enjoyed the plentiful game along the river-banks and ate many fine fish. They also noted the dangers.

> There are many piranhas, ferocious fish with terrible jaws. They destroy nets in which they are caught, and only the strongest hooks suspended from stout wire answer; others are useless. Bathing, by reason of the piranhas, is dangerous except in the rapids in shallow water. There are also electric eels and sting-rays, both unpleasant bathing companions. The principal pests in the camps were *tucandeiras*, giant ants whose bite is excessively painful, small biting flies (*piums*), mosquitoes, ticks (*carrapatos* of the species *amblyomma* and *dermacentor*, suspected of transmitting *Leishmania americana*), *mucuims* or red bugs, and chiggers. Lara spiders (*caranguejeiras*) were encountered occasionally, one the diameter of whose circle measured $10\frac{1}{2}$ inches (27 centimetres). Five *sucuriju*, anacondas or water boas, were killed on the voyage between *cachoeira* [waterfall]

> Tipurema and Kulekulaima, and of the venomous serpents the *jararacuçu* (*Lachesis lanceolatus*) was most met with, usually in the camps.

The explorers pushed up the Uraricoera for over three months, helped by Maiongong and Maku Indians. They made contact with the Yanomami or Xiriana, who proved to be perfectly friendly – not the rumoured savage warriors feared by tribes downriver. The canoes were abandoned upstream of the Auari and the expedition cut trails up into the Parima hills. It made observations westwards towards the upper Orinoco before starting back on 1 May. The descent to Boa Vista took only twenty days. The Maku showed Hamilton-Rice a route through the Maracá Channel to the south of Maracá Island. His party took three days to descend this, and were the first non-Indians to do so. They reported that:

> seven major *cachoeiras* and an infinite number of lesser, bad enough descending, form heart-rending obstacles to any party working upstream. There are also long stretches of places so narrow, water so shallow, and vegetation so dense that rocks must be prised away, canoes dragged, and passageways slashed. Towards the western part of the south channel there is a great bay and many islands. The intricate plexuses of channels coming in from everywhere and going nowhere form a *rete mirabile*, confusing and inextricable to all but an Indian . . . In no part of South America does another such river exist.

Alexander Hamilton-Rice, like Robert Schomburgk ninety years previously, published the account of his exploration of the Uraricoera in the *Journal of the Royal Geographical Society of London*. The RGS awarded Hamilton-Rice its highest royal medal – just as it had done to Schomburgk in the preceding century.

CHAPTER TWO

The Maracá Rainforest Project

Forested hills near the Purumame waterfall. (*Haraldo Palo*)

EARLY in 1984, Dr Paulo Nogueira Neto, the head of the Brazilian Environment Secretariat, SEMA, wrote to invite the Royal Geographical Society to send a team of scientists to make a comprehensive study of the Maracá Ecological Reserve. 'Our basic objective', he wrote, 'is that research should be undertaken to provide a full understanding of the ecosystems that exist there. We want to obtain a general ecological survey of the station in the following disciplines: geology, geomorphology, soil studies, climate, hydrology, vegetation and fauna.' Once such data were obtained, the Reserve could be zoned and recommendations made for future research.

This was just the sort of work that the RGS could undertake. Founded in 1830, the Society had supported all the great British explorers since that time, everyone from Dr David Livingstone to Scott of the Antarctic and John Hunt and Edmund Hillary of Mount Everest fame. In recent years the Society has organised big research projects involving teams from different disciplines. In the late 1960s, it sent such an expedition to Mato Grosso in central Brazil; and in 1977–8 it mobilised over 100 scientists to make an environmental survey of the Mount Mulu National Park in Sarawak for the Malaysian authorities. That Mulu Expedition produced a survey of a rainforest reserve that was very similar to the one now requested by Dr Nogueira.

The RGS accepted the invitation from SEMA, which later became part of Brazil's larger environmental agency, IBAMA. I decided to lead this Project myself. I had been Director of the Society for almost ten years and had helped launch a series of successful multidisciplinary projects, but had not had the challenge of leading one. I knew Brazil quite well, having been on a tough six-month exploration of the upper Iriri River with the Brazilian mapping institute, IBGE, and later spent two years on a series of expeditions and visits to tribal Indians in many parts of that vast country. I had written books on the history of Brazilian Indians and edited two volumes about recent changes in Amazonia. So I went to Brazil to meet Paulo Nogueira Neto and to visit Maracá.

Maracá is a huge riverine island in the heart of South America. It lies some 120 kilometres northwest of Boa Vista, the small capital of Brazil's northernmost state, Roraima. The island is triangular, the same shape as the British Isle of Wight but roughly twice its size. It is enclosed by two channels of the powerful Uraricoera River, a headwater of the Rio Branco which in turn flows into the Rio Negro and the main Amazon. The Amazon basin is so immense and so flat that the waters pouring past Maracá still have some 2500 kilometres to flow before entering the Atlantic Ocean, yet the island is at an altitude of only 100 metres above sea-level. Most of Maracá is covered in rainforests, but there are patches of open savanna, seasonally flooded wetlands, creeks and low hills. Although some Indians once lived on Maracá, it has never been colonised. It is now uninhabited and mostly unexplored, and has enjoyed protection under SEMA's care since the mid-1970s. As a

result, the Maracá Reserve is a haven for abundant wildlife.

SEMA's local administrator, Guttemberg Moreno de Oliveira, drove me to the Reserve – a bumpy four-hour drive across the open plains of Roraima. There were superb views over rolling grasslands interspersed with lines of graceful *buriti* palms, distant views of table-mountains, dramatic cloud formations, and very occasional clusters of cattle or isolated ranch-houses. We finally reached the crossing of the southern channel of the Uraricoera River and were ferried across by Amazonas, the Reserve's skilled and resourceful care-taker. Then there was a 2-kilometre drive on a causeway over wetlands to the ecological station.

Maracá is one of the world's best purpose-built field stations for tropical rainforest research. It consists of two long buildings fronted by verandas, and three smaller houses. One building contains the dining-room, kitchen, sitting-room and a series of dormitories with bunk beds and tiled shower-rooms. The other building has two laboratories, store-rooms and an office, and there are further sleeping-quarters in the other houses. Guttemberg and Amazonas kept Maracá functioning smoothly. But this excellent comfortable station was empty: I was amazed to find that no scientist had worked there during the previous eighteen months.

The next two years were spent planning what became known as the Maracá Rainforest Project. We wanted to find good scientists to fulfil the daunting task of compiling the ecological survey requested by SEMA. The only way to attract the best people was to offer them the chance to pursue their own research objectives, but within defined sub-programmes, and with the obligation to supply any data about their speciality to the Ecological Survey. I started to recruit rainforest experts, preferably people who had worked in Brazil and spoke Portuguese. Some of these had taken part in the RGS Mato Grosso Expedition or the more recent Mulu (Sarawak) Expedition, so I knew the quality of their work.

We also decided to include a programme of human geography, studying settlers who are actively colonising the savanna–forest boundary in northwest Roraima. This Land Development programme was of course based away from the Maracá Ecological Reserve.

We were lucky to find Dr Jim Ratter to lead the

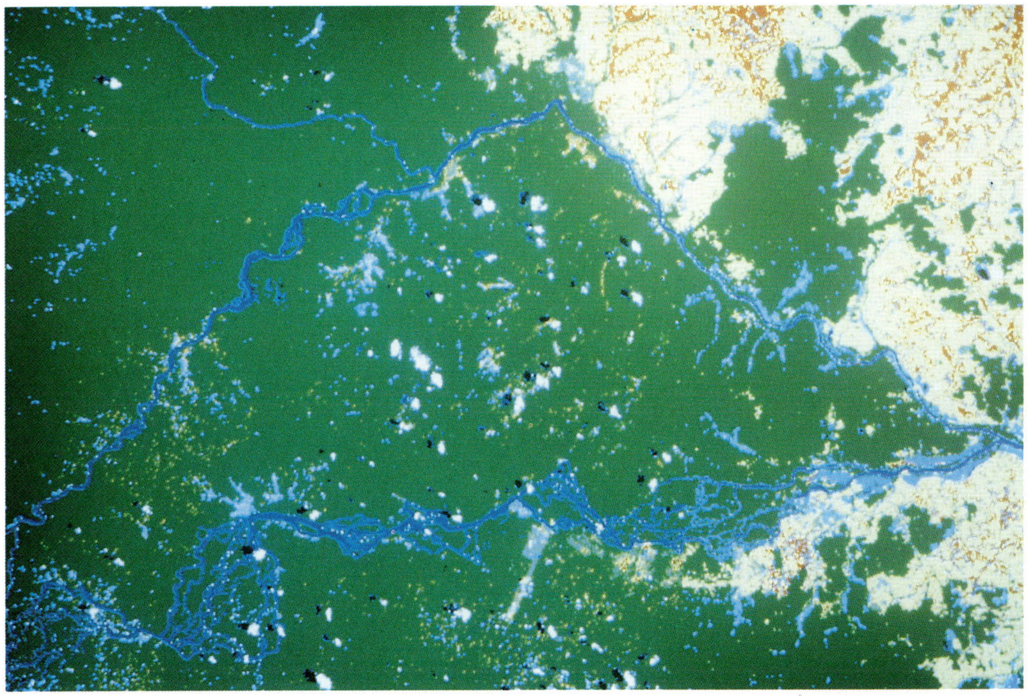

◀ Wetland and the forest–savanna boundary that fascinated scientists. (*Chris Caldicott*)

▲ A satellite image shows triangular Maracá Island covered in forest, but with natural savanna and cleared ranches to the east. (*Landsat*)

▼ The five buildings of the Maracá ecological station in their forest clearing. (*Haraldo Palo*)

Ecological Survey. Jim is a Senior Principal Scientific Officer in the Royal Botanic Garden, Edinburgh, and during the past twenty years he had gained an unrivalled knowledge of Brazilian flora, particularly of savanna plants. He speaks good Portuguese, has an endless fund of anecdotes, and mixes a famous *caipirinha* cocktail of *cachaça* rum and lime. His was the most difficult task: to extract data from a wide range of experts – botanists, entomologists, zoologists, geologists and soil scientists – and weld it into a coherent Ecological Survey. We knew that it would be impossible to do more than scratch the surface of a reserve as big as Maracá. Its forests, savannas, wetlands and rivers contain hundreds of thousands of species of organism. We hoped to attract as many specialists as possible from within Brazil to contribute data to this Survey.

Four other leaders were invited to devise research programmes: John Proctor, a study of rainforest regeneration; Victor Py-Daniel, research on insects that transmit Amazonia's worst diseases; John Thornes, experiments on the water cycle within a forest and effects on the soils of different types of clearing; and Peter Furley, investigations on the changing nature of the agricultural frontier and the problems of frontier settlers.

Each of these programme leaders visited Maracá to plan their research proposals, and various members of the Project went to the famous Amazon Research Institute, Instituto Nacional de Pesquisas da Amazônia (INPA) in Manaus, to discuss Brazilian participation. INPA is the world's leading centre for rainforest research, with a wealth of talent and experience among its hundreds of scientific staff. This is appropriate, for the Amazon basin has over half of the world's surviving tropical forests, and most of these are in Brazil.

INPA welcomed the Maracá Project and assigned Dr Ângelo dos Santos, director of its ecology division, to be in charge of liaison with the RGS and SEMA. Scientists from INPA were involved from the outset in each of the five research programmes. They participated with increasing enthusiasm and skill as the Project progressed, particularly in the Ecological Survey and Medical Entomology programme. In the course of the thirteen-month field phase no fewer than fifty-three scientists and forty-one scientific technicians from INPA worked on the Maracá Project, so that the Amazon Research Institute was by far the largest single source of scientific talent. INPA's campus in Manaus housed the main review conference, and its journal *Acta Amazonica* is publishing most of the Project's technical papers.

Although the Royal Geographical Society provides its sponsored projects with some staff, office overheads and financial underwriting, each of these ventures is expected to be financially self-sufficient. This means that it has to raise its own funds. For over a year, helped by a committee chaired by Viscount Montgomery of Alamein, I set about raising the money for this ambitious project. The main sources of funding were grant-giving research councils, charitable trusts concerned with the environment, companies – particularly British companies with interests in Brazil – and individual members' own contributions.

Fund-raising was a slow process, with many disappointments. I spent many days in Rio de Janeiro and São Paulo, visiting companies to beg for money or support in kind. Senior businessmen were generous about seeing me at short notice, often coming into their offices early or staying late to fit in an appointment. I found myself in elegant modern offices, high up in splendid corporate headquarters, but felt rather inhibited asking for funds for British scientists to study rainforests thousands of kilometres away at the farthest end of Brazil, when there was so much poverty and distress right outside the windows of the companies I was visiting. Many executives heard me out politely but declined to help. It was therefore a wonderful tonic when I called at the building of ICI Brasil and was greeted by its dynamic General Manager, John Matthews, and his enthusiastic Director of Research, Dr Galdino Vieira. Here at last were men with knowledge of the Brazilian forests, true woodsmen who appreciated the importance of what we were proposing. ICI became the Project's most dedicated backer. Apart from a cash donation, 'Dino' Vieira sent his assistant up to Roraima to see to the welfare of the Project's members. ICI paid for a mass of necessities and luxuries for the base-camp buildings, including a metal boat, a bottle-gas-powered refrigerator and even a television and video that delighted the station's administrative team. Kenneth Light of the British-owned cigarette company Souza Cruz was very supportive, with a cash donation and a lightweight motorcycle that proved invaluable for running errands around Boa Vista.

In the end, enough grant-giving agencies backed our proposals, and a number of companies also responded generously. The Project's main supporters were: The Ford Foundation, whose Brazilian bureau funded the Ecological Survey, Land Development and Forest Regeneration programmes; Tim and Susie Sainsbury's Headley Trust; the British government's Overseas Development Administration; The Baring Foundation; The Royal Society with a series of individual grants; and The Dulverton Trust. Companies who responded most enthusiastically to our conservation plans were: ICI Brasil; BAT's Brazilian subsidiary, Souza Cruz; Lloyds Bank International; and Eurobraz bank. Mercedes-Benz lent a 13-ton truck that became the main supply workhorse; Mercury lent three fine 25-hp outboard motors; Sachs-Dolmar lent chainsaws; Petrobrás provided fuel; and many others gave money or supplies (see Appendix Two).

Although SEMA had invited RGS to work at Maracá, each scientist and research programme still had to submit to painstaking vetting by the National Research Council, CNPq. This process took almost a year, and the delays caused some good scientists to leave the Maracá Project and accept invitations from other countries.

We gradually realised that the Project had run into fierce opposition from some Brazilian scientists. This was partly because of a nationalistic, xenophobic antipathy to foreigners working on Brazilian research problems. There was also suspicion about the size of the Project: Brazilian academics were not used to such a large multi-disciplinary effort, which gained a high media profile from its sheer size, and we were therefore restricted to thirty British scientists. We fell foul of inter-ministerial friction, too. Our host, SEMA, was within the Ministry of Urban Development and Environment, whereas the scientists of the research institute INPA came under the Ministry of Science and Technology. Although I and various other British scientists called on the Director of INPA, Dr Herbert Schubart, and on other department heads, news of the Maracá Project did not reach all parts of that sprawling research complex. Those who had not heard about it were resentful.

As the months dragged by without confirmation that we would receive research visas from the National Research Council, we grew increasingly desperate. His Royal Highness The Prince of Wales, who had honoured the Project by agreeing to be its Patron, had accepted a date for the launch, but we had not yet received formal permission to work in Brazil. Even though we had been invited almost three years earlier by the Brazilian Secretary of the Environment, each member of the Project still had to submit to a cumbersome screening process that culminated in approval by the National Security Council. Lord Montgomery appealed to the Minister of Science and Technology, Renato Archer; so did the British Ambassador, my friend John Ure; and so did Archer's personal friend the Brazilian Ambassador in London, Celso de Souza e Silva. The Minister finally granted approval in November 1986, just in time for the launch. I do not know whether he had to overrule objections from INPA, or short-circuit the Research Council's slow machinery, or appeal direct to the President of Brazil. We later heard rumours that all these manoeuvres had taken place on our behalf.

The Project was launched as planned at a party in the Royal Geographical Society's House. The Prince of Wales wished it well, stressing how important rainforests are to mankind, and commending our practical approach to their study and regeneration.

The political struggle was not over, however. When I was in Brazil early in 1987, raising funds and preparing for the arrival of the first team of scientists, I was told by someone in SEMA that the Forest Regeneration programme had been cancelled. The head of CNPq wrote me a formal letter confirming this cancellation. This programme was one of our most exciting and potentially valuable research initiatives. Its team had received their visas and funding for their work. Its cancellation at that late stage and for no reason was unthinkable. I hurried to Brasilia to appeal in person to the new Secretary of the Environment, Roberto Messias Franco (who had succeeded Paulo Nogueira Neto the previous summer), and to the Director of the National Research Council. Each of them said that he personally approved of this regeneration research. Despite this, I never succeeded in getting a written reinstatement of the programme.

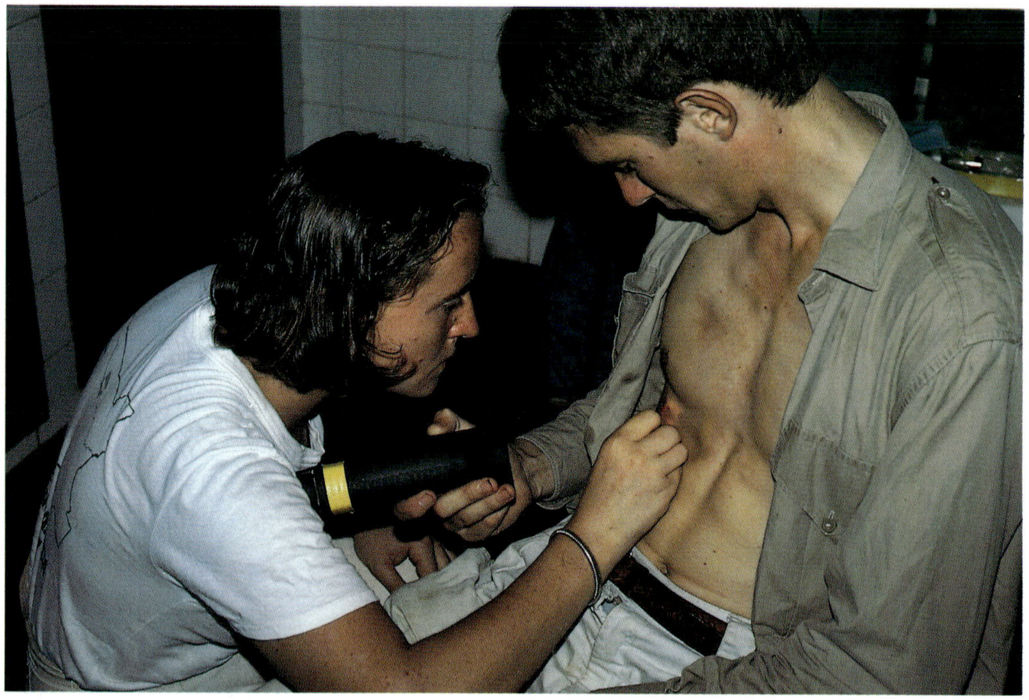

The Project's nurse Sarah Latham removes botfly larvae from Steve Bowles. She also cured his piranha bite. (*Duncan Scott*)

We also learned that the influential but politically biased Brazilian Society for the Advancement of Science sent a telex to President Sarney deploring the fact that our Project had been granted approval. This was done before our arrival and without any reference to our hosts, SEMA, or to the RGS to learn more about our plans. It was behaviour contrary to usual international courtesies between scientists, and it had no effect.

I was sure that this hostility from a few Brazilians would vanish once we were working together in the field. We would soon enjoy the usual friendship and hospitality that I had always known in that country. To smooth over differences, we tried to recruit scientists who had already worked in Brazil and who therefore loved the place and spoke Portuguese. Throughout the duration of the Project, we kept a low profile politically, never criticising local policies on rainforest management or Indian affairs. We were determined that the Project should be seen to be Brazilian. The majority of its members should come from the host country, and the results should be presented first in conferences and publications there. This was not to be a colonialist venture that removed specimens and intellectual output, exploiting the host country as a mere location for research. There was no point in undertaking all this effort if its results were not to be applied in Brazil, the country that has 30 per cent of the world's surviving tropical forests. Also, the Brazilian scientific establishment is far too large and distinguished to be bypassed. We needed to enlist as many local members as possible if the Project was to succeed and if we were to fulfil even part of the Ecological Survey requested by Dr Nogueira.

I had to create a new administration team for the Maracá Project. We found five outstanding people, hard working, well qualified, good linguists, and with the right blend of easy-going efficiency to make things work in a remote part of Brazil. The Field Leader was Steve Bowles, an environmental sciences graduate from Southampton University who had already proved his toughness and skill in man-management by leading an expedition to Amazonia. The Administrator was Fiona Watson, with a languages degree from St Andrews University, who had been instructing businessmen about Latin America. Always cheerful but firm, indefatigable and efficient, Fiona kept supplies and people moving to Maracá from the bungalow that SEMA put at our disposal in the city of Boa Vista. The health of the Project was in the capable hands of Sarah Latham, a State Registered Nurse with much tropical experience. In addition to tending a daily round of minor ailments, on several occasions Sarah had to make difficult medical decisions that saved her patients. William Milliken, who graduated in natural sciences from Cambridge, proved such an outstanding botanist and all-rounder in handling boats and penetrating forests that we added him to the salaried team as a field scientific co-ordinator. After the field phase, William worked with Jim Ratter in compiling the ecological survey. In London, Daphne Hanbury kept reports, accounts, equipment and people moving with energy and skill. INPA later assigned two administrators and drivers to the Project: José Antônio Gomes and Carlos Paredes.

SEMA's permanent staff had the task of keeping Maracá ecological station functioning smoothly despite the unprecedented influx of researchers. Guttemberg Moreno de Oliveira achieved this admirably. A former broadcaster from northeast Brazil, he was always calm and dedicated, knew everyone in Boa Vista, and had good mechanical skills. Guttemberg grappled constantly with the cantankerous generator, pumps, vehicles and other machinery. Soon after the start of the

Project the station's wells ran dry and there was a wretched month when no one could shower in the intense heat. Guttemberg solved this crisis by pumping water for over a kilometre direct from the Uraricoera River. His men kept the causeway serviceable when it was muddy and pot-holed by heavy rains. The caretaker, Amazonas, was another very competent all-rounder, and his wife Francinete maintained morale by cooking some 9000 man/days of meals during the thirteen months of the Project. There was a fluctuating group of workmen, paid by the Project to cut trails and perform a mass of other tasks. By far the most impressive of these was Luis Pestana, a forty-year-old driver and boatman. He was also a skilled woodsman, an all-rounder who did everything with calm efficiency.

The Ecological Survey absorbed the greater part of the Project's scientific effort. Eighty-nine scientists and forty-one scientific technicians worked for some 3400 days directly on this Survey. Most were specialists in specific animal or plant groups, necessary and inevitable in this area of scientific specialisation. Many were conducting detailed research on behaviour or were collecting and identifying specimens. But all were required to submit data for the jigsaw of the Ecological Survey.

Study of the natural realm of Maracá Island divides broadly into the life sciences – botany, zoology (everything from birds, mammals, reptiles and fish to earthworms and sponges, and including entomology – so important in the rainforest) – and the earth sciences (geology and pedology, or soil studies).

Maracá is on a biogeographical boundary, with open natural grasslands to the east and unbroken forests to the west. But because the Reserve is relatively accessible and because there have been some earlier botanical collections in this region, we did not expect to find a wealth of new plant species. Nevertheless, the various teams of botanists from INPA, Kew, Paraíba, Piauí and Edinburgh were well pleased. They collected some 2500 'numbers' of plants on Maracá. A 'number' is given to each individual plant collection, which usually consists of many replicate specimens. On this Project the botanists tried to obtain eight specimens of each number, to be lodged in different herbaria in Brazil and abroad. They thus gathered and pressed some 20,000 specimens.

The botanists worked from boats, collecting riverine vegetation, or on foot on transects or trail systems all over the huge island. They often took risks, climbing high into trees to catch elusive blossoms, and even using mountaineering tackle to abseil over 30 metres into the forest canopy. More often they used a *podão* – secateurs at the end of a series of long aluminium poles. They were even known to shoot off high branches or blossoms with a shotgun. The botanists returned with bales of flowers, fruits, plants and branches, and then pressed and dried their specimens.

Plant-drying was a dangerous business. At one stage, Peter Edwards of Kew dug a drying trench and, stripped to the waist in the tropical heat, kept it stoked with burning logs for days on end. On two occasions plant-drying operations almost set fire to the laboratory building: once when a cooker exploded, sending flames licking up the building's wall, and once when the lead from a gas cylinder caught fire and a disaster was averted only by brave action on the part of INPA technician José Lima dos Santos. He raced across the station to disconnect the gas cylinder, seconds before the burning lead would have caused a powerful explosion. Once dried, the plants were pressed between sheets of newspaper and sent off to their various herbaria for storage or further identification.

Individual botanists specialised in different groups or

Two scientists from INPA: Ângelo dos Santos, the Project's Brazilian counterpart, and João Ferraz examine grasses during research on the forest–savanna boundary. (*Dudu Tresca*)

▲ The botanist Maria das Graças Medina Arrais with the luxuriant flower of *Gustavia augusta*, a tree of the Brazil-nut family. (*Peter Edwards*)

▼ Gwilym Lewis of the Royal Botanic Gardens, Kew, keeping his plant press dry. (*Peter Edwards*)

► A rapid at the height of the dry season. (*Dudu Tresca*)

disciplines: Brian Stannard in parasitic Loranthaceae (mistletoes), Maria Lúcia Absy in pollen, Peter Edwards in ferns, Ray Harley in labiates (mint family), Maria das Graças Medina Arrais in bromeliads (pineapple family), Isonete de Jesus in fungi and Edileuza Sette Silva and Lauro Xavier in lichens. Gwilym Lewis, a legume (pea family) expert, collected around the entire 200-kilometre perimeter of the island and found one new species. He was particularly pleased to discover quantities of an *Alexa* with potentially important pharmaceutical properties.

In Brazil there is a type of scientific technician who is invaluable in scientific fieldwork. These *técnicos* lack academic qualifications, but their knowledge of forest organisms is unrivalled and they have the necessary skills to penetrate forest and rivers and to climb trees. Sixty-five *técnicos* worked on the Maracá Project, often for long periods or on repeated visits. Most were employed by INPA and they represented one of the Amazon Research Institute's most valuable contributions to the Project's success.

A series of highly experienced *técnicos* helped Jim Ratter and William Milliken enter the remotest unexplored parts of the island to survey their forest ecology. They found the vegetation in the western and central parts of the island to be notably different from that near the station at its eastern tip. The western forests generally grew on higher, more dissected ground than those in the east and more closely resembled the tall forests of the Rio Negro and Amazon near Manaus.

A fundamental problem for the Maracá Project was how to reach the remote parts of the island. We could not afford movement by helicopter. It would have been possible to cut a trail for 60 kilometres to the western end, but the island is hilly, and problems of carrying supplies to camps along such a lengthy path would have been daunting. The rivers were the obvious route, but they were blocked by a series of formidable rapids which became increasingly difficult to pass as the water fell during the dry season. These were the same obstacles that had caused such anguish to the early explorers. Nothing had changed since the journeys of Ricardo Franco de Almeida Serra in the 1780s, of Robert Schomburgk in 1838 or of Theodor Koch-Grünberg in 1911, except that we now had outboard motors but fewer Indians.

◄◄ A calm stretch of the often turbulent Maracá Channel. (*Dudu Tresca*)

◄ River travel: Jill Thompson flanked by Duncan Scott and John Hemming, with Luis Pestana at the helm. (*William Milliken*)

▲ A Maiongong Indian boatman carried the heavy outboard motor over the steep Purumame portage. (*William Milliken*)

In our early attempts to reach the western end of Maracá we struggled to haul the boats through the swirling waters of the rapids. Although Amazonas and Luis Pestana were excellent boatmen, it was impossible for them to see rocks beneath the mass of brown, sediment-laden water. Despite their skill and caution, the outboards' propellers often hit submerged boulders and became battered with dents and pierced with holes.

Getting through rapids was hard work, with frequent off-loading and carrying heavy loads around portages. The more agile members of a river expedition jumped out and pushed the metal boat through the dangerous current, or leaped from rock to rock and heaved on ropes. Other members of the team made their way through the forest or waded and slithered along the river's edge. Clambering over jagged submerged rocks was often painful. Shins were grazed and bruised, and people wading often slid into deep water. At times, to cross stretches of raging water, the fittest members of a team jumped across on to rocks slippery with weed and then threw a rope back for the others to come across a *corda*. This meant hanging on to the rope and allowing the current to sweep you to the far side, where you could be hauled to safety. I once missed my footing in a rapid and instantly hurtled downstream underwater, luckily into a side eddy.

To pass the largest rapids, the boats were emptied of all cargo and passengers, apart from the boatman manning the outboard and the prowman. These then had an exhilarating, dangerous dash through the whirlpools, spray and high waves where the river poured between the boulders. It took great skill and fast reflexes to keep

One of the Project's boats approaching forested hills near the western end of the island. (*William Milliken*)

William Milliken learned how to navigate through dozens of dangerous rapids.

a boat from hitting a rock or swamping if it turned athwart the current.

We finally reached the western end of Maracá Island, some 100 kilometres by river from the ecological station. Our reward was to see the mighty Purumame Falls. The explorer Robert Schomburgk saw this waterfall in 1838 and described it as one of the largest in all the Guianas. 'The grand and awful appearance of this large body of water is sublime beyond description. . . . It ultimately dashes down a precipice of 45 feet between two immense pillars of black shining rocks.' Schomburgk made his Indians cut the largest tree they could find above the falls. He watched in awe as 'it approached the chasm and with its numerous branches was hurled down the precipice. Scarcely had it approached the whirling gulf, when it was sucked under, and when it reappeared it was already in the middle of the outflowing stream, a naked trunk, its countless branches broken, and rushing down the impetuous channel.' The Indians have a portage trail bypassing the Purumame Falls, but it climbs over a granite hill so steep that you have to clamber up on all fours. To see the central part of the waterfall, we clung to trees on a vertiginous, cliff-like slope, with the Uraricoera River roaring below.

The problem of the rapids was gradually surmounted. Victor Py-Daniel's Medical Entomology team made an epic journey: its two long aluminium boats were flown to an Indian post far up the Uraricoera River and then came down it and through the labyrinth of islands, rivers and waterfalls of the Maracá Channel to the south of the island. That team made friends with the Maiongong Indians, a Carib-speaking tribe related to the Maquiritare of Venezuela. The Maiongong are lords of the Uraricoera, with an intimate knowledge of its every rock and rapid. It was a Maiongong who in 1838 told Schomburgk that he could attempt the Uraricoera route. The Maiongong manoeuvre their boats through a tangle of rocks with the aplomb of urban commuters weaving through rush-hour traffic. Approaching a raging rapid, they would ease the boat up to look at various possible routes, have a quick consultation, put the outboard on full throttle and hurtle up through the spray, zigzagging and missing rocks by a matter of centimetres. We engaged Maiongong guides. Luis Pestana learned their routes and William

◀ The botanist William Milliken beside stilt roots of the paxiuba palm (*Socratea exorrhiza*), during cutting of the Fumaça trail. (*John Hemming*)

◀◀ Victor Py-Daniel, leader of the medical entomology team, examining Simulid black flies that transmit onchocerciasis, river blindness. (*Steve Bowles*)

Milliken mapped the tortuous system through each maze of rocks. Thus, by the end of the rainy season in late 1987, we were able to send boats up to the farthest end of Maracá, and two groups even completed the tough 200-kilometre circumnavigation of the island – the first time this has been done. It involves overcoming some forty formidable rapids.

Teams moving out to penetrate unexplored parts of the enormous island established camps on the river-banks. A camp consisted of a rectangular frame of thin, straight trees. Two long horizontal beams were braced against the inward thrust of a line of hammocks slung between them. Above, pitched roof-beams supported a large tarpaulin. The diet in these camps was basic foods (rice, manioc flour, coffee and sugar) and tins, including a fine supply of corned beef from Frigorífico Anglo. This fare was enlivened by some fresh fish from the river: piranha, a bland and bony fish, tended to be the most easily caught.

On one occasion, piranha took their revenge. The field leader, Steve Bowles, was wading in a quiet stream when he was suddenly attacked. A piranha tore a nut-sized chunk from his foot, and there were other teeth-marks near the bite. Steve managed to hurry ashore and then had to pilot his boat for 5 kilometres and walk a further two to reach the ecological station. Just as he entered Sarah Latham's dispensary his wound opened a small artery and the room was covered in blood. Had he bled so profusely in the river, frenzied piranha might have savaged him. Piranhas and other dangerous Amazonian fishes are discussed further in Chapter 4 (pages 93–8).

Steve Bowles, the Project's field leader, kept the boats operational. (*Jill Thompson*)

▲ Edinaldo Nelson dos Santos Silva collecting zooplankton. (*Mark O'Shea*)

► Rogério Gribel of INPA collecting a bat. (*Duncan Scott*)

The forests near the ecological station were festooned with traps for everything from bats and insects to reptiles and rodents. Some scientists simply observed. Andrea Nunes, an attractive Brazilian primatologist, for many months spent every day alone in the forest pursuing families of monkeys as they leaped from branch to branch in the tree-tops. No one knew where she was. She strayed far from the trails and carried no compass. But she reappeared every evening, excited by discoveries of her monkeys' behaviour (see page 106).

Elton Pinto Colares from INPA made a study of otters, while others collected small mammals. Fish in the Uraricoera River were studied, and there was work on zooplankton and other aspects of limnology, with discovery of species of tiny plankton new to science.

Teams of ornithologists recorded an amazing total of well over 450 species of bird in the island's forests, swamps and savannas. Many birds were new to this part of Amazonia, and there were discoveries in their behaviour patterns. Maracá's rivers were like aviaries, alive with formations of dazzling macaws, green ibis perched on rocks, trees full of ungainly primeval hoatzin, flights of bright tiny parakeets, herons, pure white egrets and birds of prey hovering overhead. Our bat people were equally successful. They caught forty-eight species, everything from vampires to frog-eating bats, which makes Maracá richer in bats than any other known South American forest. Fif Robinson, a bat expert from Whitby, and Rogério Gribel from INPA released specimens they caught, after photographing them and measuring everything from their wingspans to their teeth structures. Fif explained: 'They breed so slowly, they do so much good, and they face so many hazards and enemies, that I would hate to reduce their numbers any further by killing and collecting them.'

Some of the Project's best scientists came from Rio Grande do Sul or Paraná, far away in southern Brazil. Professor Cecília Volkmer Ribeiro worked on sponges in the lakes and wetlands, while her friend Maria Cristina Mansur studied molluscs. They were particularly active in a grassy lake in the midst of the rainforest a few miles north-west of the ecological station. Such a lake is a rare phenomenon in a tropical forest. The one on Maracá Island is surrounded by *buriti* and *açaí* palms and is full of a luxuriant mass of aquatic vegetation, teeming with life that includes rare and curious forms of sponge. Professor Arno Lise, from the same Zoobotanical Foundation in Rio Grande do Sul, studied spiders, searching for them during both the dry and wet seasons. During the latter visit he discovered them in abundance, in a bewildering variety of types and sizes, ranging from tarantulas the size of human hands to tiny social spiders living in great colonies and collectively weaving immense webs as big as trees. Two fine butterfly collectors from Curitiba, Mirna Casagrande and Olaf Mielke, pursued their beautiful quarry with classic nets.

Insects account for over 50 per cent of the animal biomass in a tropical forest: they are present in such prodigious quantities that their combined weight far exceeds that of all other creatures. José Rafael of INPA was the Project's leading entomologist, supervising the collection of 30,000 insect specimens. The entomologists used every form of capture: mist nets, malaise nets, light traps and tree-fogging (in which a tree is gassed and its insects are collected as they fall to the ground beneath). As Dr Rafael's own speciality is horseflies, he borrowed an old horse from a neighbouring ranch, which was swum across to the island, hobbled, and left as unwilling bait for the flies. Other entomologists collected or observed other species: Pierre Jolivet worked on beetles and ants, Tony Raw on wasps, and Francisco Peralta and Father Jesus Moure on bees.

Forbes Benton, a British entomologist who has worked for many years in a Brazilian government research establishment, was on the Project studying ants and in particular fly parasites that prey on leaf-cutter *saúva* ants. He spent months on the Project, out in the forest every day and many nights observing his insects. Forbes also devised a series of experiments on pollination of flowering vines. He used chicken-wire and netting cages on flowers to see which insects robbed pollen and what vector took it to pollinate other blossoms.

Mark O'Shea, a red-bearded herpetologist, pounced on every form of reptile and amphibian and soon had a fearsome collection of snakes, lizards and poisonous frogs in one of the laboratories. If he saw movement in the grass he would immediately dive to catch its author, deciding as he leaped whether or not the snake might be venomous. He boldly plunged into murky ponds full of *jacaré* alligators, the largest of which were 2 metres long, in pursuit of his quarry. Mark also

Olaf Mielke and Mirna Casagrande, butterfly collectors from southern Brazil. (*Steve Bowles*)

▲ José Rafael, leader of the entomology team, with a butterfly net, insect trap and a hobbled horse used as bait for biting flies. (*Duncan Scott*)

◀ Mark O'Shea with a *papa-ovo* ('egg-sucker') cribo snake (*Drymarchon corais corais*). (*Mark O'Shea*)

The herpetologist Mark O'Shea sexing a *jiboia* boa constrictor (*Luis Claudio Mango*)

used drift traps, which consisted of a long fence of knee-high green canvas stretched across the forest floor. Plastic buckets were sunk into the ground at intervals under this fence. Small amphibians and other creatures would encounter the canvas barrier and run along it until they fell into one of the buckets, and each morning Mark went out to collect the night's haul from his traps.

Mark also took pictures of his specimens. One day he was photographing a young rattlesnake when he was disturbed by someone entering the laboratory behind him. He slightly relaxed his grip on the reptile's neck and it struck his other hand holding the camera. The bite was potentially lethal. Mark was saved by the Project's nurse, Sarah Latham, who correctly applied serum and nursed her patient through a dangerous night. Mark was evacuated by plane next day to hospital in Boa Vista.

When the thirteen-month field phase of the Maracá Project ended in March 1988, 203 scientists and scientific technicians had worked for a total of over 6000 days. This total breaks down as 117 Brazilian scientists, 31 British scientists and 55 technicians, but as the British members generally spent more time with the Project, the number of days worked by Brazilians and non-Brazilians was roughly equal. This quantity of researchers makes the Maracá Project one of the largest ever mounted in Amazonia by any European country, and certainly the biggest of its kind in Roraima.

In addition, over 3000 man/days of food was consumed by the administrative team, which included SEMA's people and various local workmen employed on trail-cutting, students and teachers, press including a television crew, and other visitors.

The effort of feeding and supplying so many people was daunting. The administrative team worked gallantly and tirelessly to keep people, food, drink, bottled gas, fuel, instruments, scientific supplies and a mass of other cargo moving to and from Maracá Island. Although Boa Vista is a spacious and attractive small city, almost all its supplies have to come 850 kilometres by river or dirt road from Manaus. There were therefore times when such basic commodities as beer or gas were unavailable.

Not surprisingly, morale dropped whenever Maracá ran out of cold beer. It would be sold only in exchange for returned empty bottles, so there was constant effort heaving full and empty cases on to and off pickups, canoes and the lorry, and up and down the slippery river banks. At times of shortage, we had to learn about the approach of a beer truck from Manaus and be sure to be at the depot when it arrived. The same applied to heavy gas canisters, on which we depended for cooking, refrigeration and the operation of some laboratory instruments. There were many hours of queuing in the midday heat (40°C or over 100°F) to be allowed to buy just one bottle of gas per person. There was also continual movement of diesel spirit for the station's generator, which ran for a few hours every evening, and petrol for the outboard motors.

Steve Bowles and Fiona Watson toiled in the equatorial heat to buy, dispatch and load many tons of cargo,

and to organise travel arrangements and formalities for so many scientists, some of whom could be frustratingly absent-minded. The success of their efforts was demonstrated in the Project's high morale. There was a happy atmosphere at the Maracá ecological station, with occasional dances and parties, picnics on the river, lectures by researchers, and conversations every evening on the veranda of the main dormitory building. Such discussions between different scientists are one of the most stimulating aspects of a multi-disciplinary expedition.

The Maracá station is 120 kilometres northwest of Boa Vista, at the end of a dirt road that becomes increasingly rough as it approaches the island. There are stretches of severe erosion and pot-holing. The road crosses a dozen bridges and some of these were in appalling condition. The worst was the bridge over the Igarapé Grande, a fair-sized tributary of the Uraricoera: a 100-metre trellis of rickety, termite-ridden posts with a floor of broken, rotted or missing boards. We said silent prayers each time the heavily laden lorry crept across this creaking contraption. The logistic team drove back and forth on this punishing four-hour journey many times each week and hundreds of times in the course of the Project. The mighty 13-ton truck lent to us by Mercedes-Benz was the main work-horse, moving great quantities of cargo and people. We also had two battered Toyota four-wheel-drive pickups, one belonging to SEMA and one lent by the Mineração Vale do São João mining company.

Once the constant mass of food, supplies, scientific instruments, people and luggage had reached the Uraricoera River at the end of the bumpy run from Boa Vista, everything had to be manhandled down a slippery bank into canoes. Countless man hours were spent on this labour, hauling loads down into the boats and then unloading them on the muddy far bank, for transport along the two-kilometre road to the station.

On one occasion, when skilled driver Luis Pestana was taking a Toyota along the main road out of Boa Vista, its back axle snapped. The vehicle went out of control and somersaulted. Passengers in the cab suffered minor injuries, as did some who were thrown from the back. But the primatologist Andréa Nunes had a badly broken arm, and Steve Bowles, who was trapped under the vehicle, was severely injured, with a deep gash on his head and ear, a lung collapsed from the weight of the Toyota, and bad burns from its exhaust. A hospital in Boa Vista gave rudimentary treatment but was incapable of providing the necessary surgery. Our friends in ICI Brasil came to the rescue. John Matthews and Galdino Vieira generously authorised the prompt evacuation of both patients in specially equipped Learjets: Andrea to Rio de Janeiro and Steve to the finest private hospital in São Paulo. The Project's home agent, Daphne Hanbury, was to be rewarded by visiting the project in the field. No sooner had she reached the island than the accident occurred. She flew to São Paulo with Steve and tended him while his head and ear were operated on by one of the world's best plastic surgeons, and his burns were eventually healed by skin grafts at the Radcliffe Infirmary in Oxford. Steve Bowles and

▲ A four-wheel-drive truck lent to the Project by Mercedes-Benz carried great quantities of stores, equipment and people over the savanna between Boa Vista and Maracá. (*John Hemming*)

◀ A 'jeg' with an old Volkswagen engine transported people over the two-kilometre track from the river to the ecological station. (*Jill Thompson*)

The forest-regeneration team, Jill Thompson, Daphne Hanbury and Duncan Scott, surveying an artificial clearing. (*Chris Caldicott*)

Andrea Nunes owe their recoveries to the swift action and great generosity of ICI.

Trails were cut inland from the riverside camps. An important path was opened due south from the Fumaça rapid, some 30 kilometres west of the ecological station. This path ran over hills, through swampy areas and different types of vegetation, towards the higher ground in the centre of the island. Hacking forward on this and other trails, it was stimulating to know that every step was into unexplored territory that had probably never been seen or trodden by any man. The Fumaça trail eventually ran for 18 kilometres, almost completely across the island from the Santa Rosa to the Maracá Channel. Another trail pushed eastwards from the mighty Purumame Falls at the farthest extremity of Maracá Island. This path was uneven, moving very steeply up and down over the range of rocky outcrops that formed the waterfall. Closer to the ecological station, another trail crossed the island from north to south: we called this the sloth trail because a sloth was found on it and because the man cutting the path moved painfully slowly. Throughout the duration of the Project, a network of trails was cut or reopened close to the station. These were the paths in constant daily use by the majority of the researchers.

Cutting trails was hot and exhausting work, but strangely gratifying. A team of two or three people made the path, but the lead man did most of the hard work, slashing bushes, saplings and creepers with forehand and backhand cuts of his machete. Those behind improved the trail by cutting more undergrowth and pushing the felled vegetation aside. Hands and clothing inevitably became covered in scratches, ticks and thorns. We tried to keep the trails (*picadas*) as straight as possible with regular compass checks, and the key to successful cutting was to pause quite frequently to hone the machete blade to razor-sharpness with a whetstone. Cutting parties carried some food for lunch and possibly water as well, although they always hoped to meet a stream from which to quench the fierce thirst generated by such sweaty work. It was impossible to observe much wildlife during such a noisy activity. Although a trail cutter grew to know exactly how hard and where to strike each type of tree: young palms were easy, toppling from a single blow, but some saplings were tough as wire and sprang back from repeated slashes, *tucumá* palms were covered in wicked spines, and many lianas danced away frustratingly when struck. Skilled woodsmen such as Luis Pestana were always on the look-out for hives of dreaded hornets. When he saw one, the trail made a detour around it and everyone stopped talking and tried to cut and move as silently as possible, warily watching the hornets buzzing at the entrance to their small grey home, which hung inconspicuously under a low branch. Once, when we stopped on a riverbank and started to clear space for a camp, we disturbed a colony of aggressive 'killer' bees. The small black insects flew straight at each man's eyes, and stung. Within moments, everyone in the team had painfully swollen eyelids and was clambering as fast as possible into the boat to seek another campsite.

The trail that moved south across the island from the Fumaça rapids had to cross a large stream a short distance inland. Two large trees had conveniently fallen across this obstacle, on top of one another, and we used them and their branches as a bridge. One evening, a group returning to camp made its way across the two trees. I was the last of this group and, as I crossed the tangle of tree-trunks and branches, the upper tree collapsed and I fell with it. For an instant, I was suspended by a branch hooked under my rib-cage. I managed to

crawl back on the lower tree without falling into the river, but the pain was intense for many days.

The Maracá Project's four research programmes in addition to the Ecological Survey were designed to learn more about the mechanisms of a rainforest and about the people moving into this frontier region of Brazil. Food and drink are pumped through a tropical forest by two processes: the nutrient and water cycles. Rainwater pours down tree-trunks or drips from the leafy canopy, and is largely captured by the growing vegetation. So are nutrients from dead matter. Dead leaves, plant and animal litter are falling constantly in this evergreen world, and they decompose fast in the heat and humidity. Fungi, bacteria and all-important termites break down decaying matter and release its nutrients, which are quickly recaptured by an almost-solid mat of roots of the living biomass. Virtually nothing escapes into the weak sandy or acidic soils.

One of the five main research programmes of our Project was concerned with the hydrological cycle, particularly the relation between water and soils. This Soils and Hydrology team carried out a number of experi-

A group setting out along the trail cut near the Fumaça rapids. This fallen-tree bridge later collapsed. (*John Hemming*)

► Duncan Scott regularly collected litter in a series of litter-tables. (*Luis Claudio Mango*)

►► João Ferraz (*left*) examines a soil trench across the forest–savanna boundary. (*Jill Thompson*)

ments devised by Professor John Thornes of the University of Bristol. It worked on three plots on a slope in *terra firme* ('firm land' or non-flooded) forest, half an hour's walk from the ecological station. One plot was left undisturbed, as a control. Another had the forest canopy removed, but it still contained low vegetation. The third was completely cleared of both vegetation and the all-important root-and-litter mat on the forest floor.

Within these three plots, the geographers (particularly Sheila Ross and John Thornes) studied four processes: soil hydrology, which means the movement of water on the surface and below ground; soil erosion; decomposition of leaf litter; and soil chemistry. They used a wide range of testing equipment. There was a wicked-looking radioactive neutron probe for measuring sub-surface water, tensionometers augered to different depths, lengths of tubing hammered into the ground for cores, and metal troughs at the base of the slope to collect eroded soil and water run-off. As with so many of the scientists, this team was prodigal in the time and effort spent gathering data. Sheila Ross was out in the forest all day on the hot, sweaty work of conducting the experiments, and then laboured late into the night in the station's laboratories and then back in Bristol.

As has been explained, Maracá Island lies on a great biogeographical boundary. To the east are plains of open savanna grassland, stretching towards Guyana and Venezuela; to the west, unbroken forests extend across the Parima hills to the upper Orinoco. The scientists were fascinated by this transition from forest to savanna. Experiments were rigged up in pits and trenches across the forest–savanna boundary. The water-table was measured regularly, but it rose and fell like a yo-yo with each rainfall. João Ferraz and Flávio Luizão of INPA analysed soil composition, chemistry and dynamics under the forest and savanna. A team under Lucille Antony looked at soil micro-organisms, while Jim Ratter and John Proctor recorded the competing types of vegetation.

John Proctor of Stirling University devised brilliant experiments to unlock secrets of forest regeneration. If his team could discover how a rainforest grows again naturally, this would help reverse the destruction we deplore. Nothing in nature is simple, least of all in a luxuriant tropical environment where each tree species has to be scattered to escape its parasites and predators. Sadly, you cannot recreate a rainforest with a single type of tree, as in temperate reforestation with conifers. A rainforest consists of hundreds of species interrelated in a bewilderingly complex ecosystem. Each type of tree has evolved different mechanisms for pollination and seed dispersal. They produce blossoms, fruits and nuts to appeal to different insects, birds, rodents, monkeys or bats.

Proctor had a series of plots cleared, bringing trees and creepers crashing down in the interest of science. Before this felling, over 9000 trees had been tagged, measured and identified by Jim Ratter, William Milliken and the botanical technicians. It was a luxury to have such skilled botanists present. It is very difficult to

◀ Jill Thompson recorded the growth of thousands of seedlings for the forest-regeneration programme. (*Fiona Watson*)

◀◀ Jim Ratter, leader of the ecological survey and an authority on Brazilian flora. (*Fiona Watson*)

identify trees in a rainforest since their branches are high in the canopy and they blossom at different times. Normally, John Proctor would have had to remove many samples for identification elsewhere.

Twelve artificial gaps were created, covering a total of 1.5 hectares, and three identical control plots were left undisturbed alongside them. We contemplated setting fire to these clearings so that the experiments would approximate regeneration in typical slash-and-burn farming. But it was late in the dry season. The forest was dangerously dry and we feared that fires might burn out of our control. The gaps were carefully surveyed and marked out with a grid of measuring-tapes. Botanists and entomologists took advantage of the felling to collect from the fallen canopy.

Throughout the next twelve months, Jill Thompson and Duncan Scott of the Forest Regeneration team measured everything that happened as the forest grew back. They were particularly interested in the dynamics of seedling growth, and the sequence in which different species established themselves. Six thousand seedlings were individually tagged with metal number-plates, then regularly measured to see which thrived and which died. Meanwhile, litter was painstakingly collected at intervals from a series of seventy-two table-like hardwood traps that Steve Bowles got a carpenter in Boa Vista to make. This was to test rates of decomposition. The nutrient content and rate of nitrogen mineralisation of soils in the plots were measured. So also was the emergence of seeds from sixty-four trays buried in the forest floor. Another experiment looked at the size and nutrient content of leaves at different heights on the trees.

Jill Thompson kept a journal of her months on the Project. When she first arrived, she was warned of the danger from peccaries, boar-like animals that could attack human beings when they were in large packs or with their piglets. Jill recalled her fear when she first heard the clacking of peccaries' teeth and smelled their strong odour.

> Thinking the peccaries were about to launch an attack, we took to the trees. My first attempt was a liana which promptly sank to the ground, followed by my forcing Rab [Marrs] further up his rotting branch in a panic. My knees were trembling so much the whole tree shook, and my heart was doing 99 miles an hour.

The soil scientist, Dr Sheila Ross, recalled a similar predicament:

> The main problem about climbing a tree to escape from peccaries is that you leap in a frantic state up the tree to a moderately secure position, wedging your feet into the forks of the branches. Then you try to relax a little, to let your pulse-rate calm down and to brush the bark and leaves and muck out of your eyes and hair. Then, as you're beginning to settle back to normal, you realise that your problems have only started, as whole armies of termites and leaf-cutter ants start swarming over your hands and up your shirt-sleeves, from your now not-so-firm clutch on the bark of the tree.

Sheila Ross and her helper Eronias dos Santos Oliveira taking recordings from instruments in a soil plot from which vegetation has been removed. (*John Hemming*)

▲ A boat on the Santa Rosa channel, with Tepequém table mountain in the distance. (*William Milliken*)

◀ The Project leader, John Hemming, on rocks below the Purumame falls that have inscriptions carved by earlier explorers: Robert Schomburgk (1838), Alexander Hamilton-Rice (1925) and Cândido Rondon (1927). (*John Hemming*)

Later in the Project we obtained some bangers from a firework shop in Manaus and planned to throw these at the peccaries if they became aggressive. But, despite occasional treeing of scientists, the peccary never posed a real threat.

The Forest Regeneration team planned its experiments with the care of climbers plotting a route up an unscaled peak. Science is less glamorous than mountaineering, but these experiments proved as tough, back-breaking and demanding as a climb, with months of sweaty work, forays up blind alleys, and occasional successes and surprises. Maracá was unusually hot, with temperatures in the forest gaps ranging from 19°C (66°F) at night to a scorching 46°C (115°F) in the shade by day. Jill Thompson and Duncan Scott worked in these hot clearings every day for thirteen months, almost without a break. Breakfast in the Maracá base-camp was early. Almost every scientist was soon off along the trails into the forest, and most did not bother to return for lunch.

After the Forest Regeneration team returned from Brazil, laboratory work continued on its vast accumulation of samples at the University of Stirling in Scotland, financed by a grant from the Overseas Development Administration. Regeneration is a slow and complex process. The team therefore went back in later years to monitor its artificial gaps on Maracá Island, and it will continue to do so for many years to come.

Conservation is not against man's interests: it should be for sustainable human development. So we mounted two research programmes to help poor settlers at the forest edge. The forest–savanna boundary north and south of Maracá Island is one of the world's last colonisation frontiers. Settlers are starting to pour into Roraima from other parts of Brazil, and they often try to carve out farms or cattle ranches from the edge of virgin forests.

I set the scene with a historical study of settlement of Roraima from the eighteenth century to the present (see Chapter 1 and pages 116–17). The population has grown dramatically with the creation of Rio Branco (later renamed Roraima) as a federal territory in 1943, government-sponsored agricultural colonies, and the opening of a road-link to Manaus in 1977. Studies were also made of the current agricultural colonies near Boa Vista, and the ways in which that city is supplied with horticulture and wetland rice.

Luc Mougeot and Philippe Léna from the University of Pará in Belém conducted interviews with settlers in a

▲ The eastern tip of Maracá Island between the Santa Rosa and Maracá Channels of the Uraricoera River. In the foreground, a rancher has cleared and cultivated forest land; in the distance lies open savanna. (*Haraldo Palo*)

► Michael Eden and Duncan McGregor of London University study soil degradation on land cleared by ranchers. (*John Hemming*)

format they had perfected in other frontier regions of Brazil. These two human-geographers took one of the Project's Toyota Bandeirantes to three locations: to Vila Brasil, an agricultural settlement in rolling open plains interspersed with islands of high forest; to Taiano, an agricultural colony founded in 1953 on fairly fertile soils beneath semi-deciduous forest; and to Tepequém, a former mining community carved out of forests on the Tepequém table-mountain, a southern outcrop of the geologically ancient Guiana Shield. They questioned fifty settlers, seeking to learn about the colonists' background, problems and successes, and the efficacy of government aid programmes. Meanwhile, Mike Eden and Duncan MacGregor of London University measured soils on fields that farmers were trying to convert from forest to cattle pasture. A Brazilian economist, Nelson Vieira, investigated the economics of these ventures. Their findings will help identify factors that usually condemn such operations to failure. This team worked out in the open, shaded only by their broad-brimmed straw hats. Ranchers would tell them when they had cleared each field, so that they collected soil samples from areas that had been deforested from one to ten years previously. They were sometimes researching in the fields of *colonião* grass, which can grow to shoulder-height so that cattle grazing in it are lost to view.

The rate of deforestation along the length of the frontier was assessed by Peter Furley (the leader of this Land Development programme) and Tom Dargie, using Landsat satellite imagery. In order to interpret the imagery, Dargie made 'ground-truth' observations, studying the terrain and vegetation in a variety of locations (often on foot or by bicycle), to provide reference points for reading the colour combinations in the imagery. All this research should be of benefit to the settlers and to the government agencies that help them. By steering colonists towards areas where they are more likely to succeed, the environmental damage they cause is reduced.

Human settlement in Amazonia is inhibited by tropical diseases. All the worst of these are spread by insect vectors. The Project therefore included a Medical Entomology programme to learn more about these insects with a view to their control or eradication. Almost the entire Medical Entomology team came from INPA and it was led by Victor Py-Daniel, a fine researcher and a tough and experienced expeditioner.

Victor Py-Daniel's team made a series of daring and arduous river journeys, rivalling those of the early explorers. They descended the upper Uraricoera and the rapid-infested Maracá Channel, went up the Uraricaá River, investigated the Amajari and descended the Uraricoera from Maracá to Boa Vista. During these long journeys they collected insect disease vectors and also surveyed the geology, limnology and other aspects of the insects' habitats.

Dr Py-Daniel himself studied black-fly (Simuliidae), the carrier of onchocerciasis river blindness that has recently been introduced into Amazonia from Africa. This disease, which is particularly prevalent among Indians, may have been brought unwittingly by missionaries who had previously worked in Africa. Studies were therefore conducted among the Xiriana (related to the Yanomami) and the Carib-speaking Maiongong of the upper river.

By far the greatest scourge of the tropics is malaria. Dr Wanderli Tadei collected its vector, Anopheline mosquitoes, using his own skin as bait and a plucked hen as a food source to keep the captured mosquitoes alive. There were a few cases of malaria during the course of the Project, but luckily all were contracted far from the ecological station, which remained free of the disease. I caught vivax malaria on an early river trip on the Santa Rosa Channel from gold prospectors who occasionally went up this river and infected its mosquitoes. Some research was also done on sand-flies that transmit leishmaniasis, and on the dreaded *barbeiro* bugs that sometimes carry the incurable and usually fatal Chagas disease. A number of *barbeiros* were found in the station's sleeping-quarters, and people were bitten by them, but mercifully the bugs proved to be uninfected by Chagas disease.

It was a sad day when the administrative team, nurse Sarah Latham and the last of the scientists left the Maracá ecological station. The Project had been an increasingly happy one, and close friendships had developed during the thirteen-month field phase. Towards the end, we organised a workshop in Boa Vista, so that the different teams could describe their work and we could all agree on a policy for presentation and publication of the finished results, which are discussed in the final chapter.

Sarah Latham, José Ramos and Cosme Damião camp in a forest shelter roofed with *Geonoma* leaves. (*William Milliken*)

CHAPTER THREE

The Land Environment

◀ Riverine forest. (*Chris Caldicott*)

As already mentioned in the previous chapter, Maracá Island lies at the junction of the great Amazonian forest, which covers 6 million square kilometres, and the large area of savannas extending for 54,000 square kilometres into Venezuela and Guyana. However, although extensive savannas are found on the mainland nearby, more than 90 per cent of the island is forested, and it is only at its eastern end that there are a few patches of true savanna. Forest is therefore the dominant vegetation. The following pages cover the plants and animals of the terrestrial and semi-terrestrial habitats; those of aquatic habitats are considered in the next chapter.

VEGETATION

The botanical survey teams spent the greater part of the year making vegetation inventories, principally in the forest, to provide a detailed description on which future conservation and management strategy can be based. The teams consisted of one or two botanists, usually accompanied by two INPA *técnicos* and sometimes one or two local helpers. The *técnicos* have an immense knowledge of Amazonian vegetation, built up over many years of experience, which was invaluable in the surveys. They were also responsible for the preparation of dried specimens. The fieldwork consisted of identifying and collecting trees and other plants, a subject which involved discussion and often some disagreement within the group, and the continual drudgery of measuring distances and trunk diameters to provide the data for calculation of densities, frequencies, basal areas per hectare, and so on. Specimens often had to be collected from tall trees and this involved difficult climbs. The *técnicos* were specialists in climbing, but some of the botanists also showed great prowess – particularly William Milliken, who demonstrated that being more than 6½ feet tall could be a positive advantage in such work. In addition to the survey teams, botanical specialists from the Royal Botanic Gardens, Kew, and from a number of Brazilian institutes made intensive collections of specific groups of plants; these included Leguminosae (pea family), Bromeliaceae (pineapple family), Labiatae (mint family), Loranthaceae (mistletoes), ferns, lichens and aquatic plants. Other botanical groups worked on forest regeneration, mineral recycling, and the interrelationship of soils and vegetation. The following pages relate some of the botanists' observations and attempt to give a feeling for the habitat.

A good place to start is on the network of little trails beside the ecological station. Here one sees that virgin tropical forest bears little resemblance to the 'green hell' of popular imagination. There is a continuous canopy made up of the crowns of tall trees 20–30 metres in height, with occasional emergents reaching about 40 metres. Below this are lower trees consisting of young specimens of the canopy species and others

Heliconia chartacea, a humming-bird pollinated herb of the forest margin.

which belong to species especially adapted to the forest understorey.* The trees cast a deep shade and one is struck by the pleasant coolness of the forest in contrast to the scorching heat of the nearby station clearing. In fact interception of light by the massed leaves of the trees is so great that probably only 1–3 per cent actually reaches the forest floor: the photographer is surprised to find that, without flash, he has to use exposures of $\frac{1}{15}$ second and full aperture with a film that gave $\frac{1}{250}$ at f8 in the clearing! Because of the dense shade, vegetation on the floor is sparse, and it is therefore easy to walk through the forest. This is almost always the case in virgin rainforest: the impenetrable jungle, where one has to hack through the vegetation with a machete, can only occur where there is enough light reaching the forest floor to allow exuberant development of vegetation, as in secondary growth after the destruction of the trees of the original forest. Trunks are slender in

* Many interesting studies have been made of the vegetation stratification of tropical forests, showing that some of the more structurally complex have five or more layers, with a number of distinct tree and shrub layers (or storeys) below the canopy. Often it is only possible to discern the existence of the ordered layering by detailed measurement of many sample areas and the preparation of profile diagrams – a more superficial examination shows the canopy clearly but just gives the impression of a chaos of trees of varying heights below it. Maximum interception of incident light is brought about by this multi-layered structure: much that escapes one layer is caught by the next.

◀ Buttress roots support a tall forest tree. (*Dudu Tresca*)

◀◀ The kapok tree (*Ceiba pentandra*) towers over the forest canopy. (*Dudu Tresca*)

relation to the heights of the trees and it is rare to find one reaching 1 metre in diameter at breast-height. However, some enormous trees do occur, particularly in the silk-cotton tree family (Bombacaceae) which includes two huge species of *barriguda* ('big belly'). These two species, the kapok tree (*Ceiba pentandra*) and the *cedro doce* (*Bombacopsis quinata*), have swollen trunks reaching their maximum breadth (sometimes 2.5 metres in diameter) up to 4 metres above ground level – they are the New World equivalents of the African baobabs. Other very large trees belong to the Brazil nut, pea and mango families (Lecythidaceae, Leguminosae and Anacardiaceae). A feature of many of the trees, and particularly of the taller species, is the presence of buttress roots. These roots project out from the base of the trunk as more or less flat, triangular, woody plates, sometimes extending to 3 metres or more up the trunk. They resemble the fins of a space rocket standing on its launching pad, or the base of some types of coat-stand. There are a number of theories concerning the function of buttresses, but the weight of evidence seems to support the most obvious interpretation: that they act, as in the coat-stand, to give stability to a tall, unstable structure. In this connection, John C. Kricher, in his delightful book, *A Neotropical Companion*, relates the story of a group of scientists in the rainforest discussing various recondite theories of the function of buttress roots. Their Indian guide listened with interest and then told them that buttresses made trees stand up. On hearing their doubts about this he went over to a small tree, cut its buttress roots one by one and then pushed it over by hand! Buttress roots are called *raizes tabulares* ('table-

Stilt roots of the *paxiuba* palm (*Socratea exorrhiza*). (*James Ratter*)

roots') in Portuguese and *sapopemas* ('flat roots') in Tupi (see page 46). They were, and sometimes still are, used by Indians and Brazilian country folk to make canoe-paddles and table-tops. Prop- or stilt-roots are other types of unusual root found in tropical trees. They are particularly common in trees of damp, muddy habitats and occur in extreme form in mangroves. They are relatively slender, cylindrical, woody roots which radiate from the lower part of the trunk and produce a multiple attachment in the soft soil. The most striking stilt-roots seen on Maracá are those of the *paxiuba* palm (*Socratea exorrhiza*), shown in the photograph above, but they also occur in plants such as the Moraceous cecropias which are abundant on disturbed ground.

Barks are generally smooth, but often in older trees have characteristic patterns of fissures or tubercles which are useful in identifying species. Such smooth barks are in marked contrast to the trees of the savannas and deciduous forests of Central Brazil where thick, corky, fire-resistant barks are the rule. Some species have aromatic resins in the bark (for example, *Hymenaea* and Burseraceae), while others have milky-white, coffee-coloured, or yellow latex (Apocynaceae, Euphorbiaceae, Moraceae, Sapotaceae). These latexes probably serve the tree as chemical defences against herbivores, but some are also useful to man: a species of rubber tree (*Hevea guianensis*) occurs at the west end of the island, as does a cow tree producing edible milk (*Couma macrocarpa*, Apocynaceae).* The colour, pattern and smell of the bark-blaze (a superficial slash made with a machete) are diagnostic for many species and are used in identification. Barks are often dappled with

◀ The *tucumá* palm (*Astrocaryum aculeatum*) has a spiny trunk. (*Dudu Tresca*)

◀◀ Young *Mauritia flexuosa* palms along the forest margin. (*Dudu Tresca*)

attractive pale green, yellow, orange or pink patterns produced by various species of encrusting lichens.

Leaves of rainforest trees are often described as being leathery, laurel-like and very uniform, but in fact there is much variation which is useful in species identification. The forest botanist therefore spends a lot of time scanning the tree-crowns with binoculars, or, at the other extreme, scrabbling around on the forest floor to discover fallen leaves. The Maracá forest is predominantly evergreen, but a number of species are exceptions and lose all their leaves during the height of the dry season. This deciduousness occurs particularly in the

* There are a number of species of cow tree in tropical South America, belonging to such diverse families as the Apocynaceae, Moraceae and Sapotaceae. The Yorkshire botanist Richard Spruce, who travelled extensively on the Amazon in the middle years of the last century, experimented with the Sapotaceous Maceranduba, the species used in eastern Amazonia. He tried the milk fresh from the tree and in coffee and reported, 'Its consistency is that of good cream, and its taste perfectly creamy and agreeable.' However, he also warned, 'It is extremely viscid . . . a property which renders it an excellent substitute for glue, but a rather unsafe article of diet, and serious cases of constipation have resulted from its being partaken of too freely.' The milk of the Maracá cow tree produces the same effect, and we were told of a *garimpeiro* (gold-prospector) who after all else had failed was finally driven to drink mercury as the ultimate laxative. Apparently this drastic treatment was successful, although one dreads to think of its long-term neurological effects.

The *sororoca* (*Phenakospermum guyanense*) is an Amazonian relative of the banana. (*James Ratter*)

emergent trees such as *Ceiba pentandra*, *Bombacopsis quinata* and the tall leguminous enterolobiums.

In general the forest vegetation is unarmed: the plants depend on distasteful substances present in their leaves to discourage unwelcome herbivores, rather than on more obvious spines and thorns. There are, however, spectacular exceptions in some of the palms. One of these, the *tucumá* (*Astrocaryum aculeatum*), sometimes attains heights of 20 metres and has black 8-centimetre spines along its great 4-metre fronds and in a series of bands ascending the trunk. Another, the *jacitara* ('devil's thorns', *Desmoncus polyacanthos*), a spiny climbing palm found on the forest margin, has the terminal leaflets of the frond replaced by brutal recurved grapnels. There are relatively few species of palm-tree in the forest, but two of the largest, the *tucumá* and the *najá* (*Maximiliana maripa*), are very abundant, tall, understorey trees, generally about 15 metres in height.

The vernacular names of plants and animals always fascinate me. In Brazil, the majority of the names come from Tupi, a widespread Indian language which is still spoken in parts of Amazonia, and which in a closely related form called Guaraní is one of the official languages of Paraguay. Tupi is perhaps the richest known native language in names of plants and animals. The Indians had a very precise vocabulary to describe the natural world about them, and their tradition and vocabulary persist amongst the Brazilian backwoodsmen, who are almost always of mixed blood. An extremely limited knowledge of a few words allows the meaning of many vernacular names to be understood: for instance, *iba* or *uba* means tree, and the *pindaíba* is the 'fishing tree' (so called for its slender, straight branches, suitable for making rods); the *itaúba* is the 'stone tree' (because of its rock-hard wood); the *carapanaúba* is the 'mosquito tree' (because it has hollows in its bark holding pockets of water in which mosquitoes breed), and so forth. Many of the plants have Portuguese in addition to Tupi names, for example the *Tabebuia* tree is called either *ipê* (Tupi) or *pau de arco* (Portuguese, 'bow tree'). Surprisingly, Brazilian backwoodsmen usually do not understand any Tupi and generally respond with surprise and amusement when the meanings of the names they use every day are explained to them.

The lowermost strata of the forest are made up of true understorey species, adapted to passing all their life in the deep shade, together with an abundance of saplings of taller species. The most striking of these understorey species is the *sororoca* (*Phenakospermum guyanense*), a banana-like plant reaching 6 metres in height. This exotic-looking relative of the banana is widespread in Amazonia, and in fact the Brazilian botanist Velloso considers its presence as a sure indicator that vegetation belongs to the Amazonian phytogeographic province. Because of the darkness of the forest floor there are remarkably few species of low-growing herbs. Among the most abundant of those that do occur are members of the arrowroot family (Marantaceae) with large, variegated and patterned leaves which make them popular as cultivated house-plants.

The trees are pollinated by animals (insects, humming-birds and bats), in contrast to the majority of

◀ Variegated leaves of a juvenile Swiss-cheese plant (*Monstera dubia*) cling symmetrically to a tree trunk. (*William Milliken*)

◀◀ The *técnico* Cosme Damião Alves da Mota climbs a slender tree to collect from the forest canopy. Stems and perforated leaves of a Swiss-cheese plant climb the larger tree on the left. (*James Ratter*)

larger temperate trees which are wind-pollinated. This means that the flowers must attract animals by means of colour and scent. However, the flowers are not necessarily spectacular – in fact many are pollinated by very small insects such as sweat-bees and flies, and are therefore very small themselves. But large, brightly coloured flowers do occur in many species. They are particularly characteristic of the Bignoniaceae (*Jacaranda* family): the tall, long-boled *Jacaranda copaia*, which sometimes reaches 45 metres in height, has conspicuous terminal inflorescences of purple trumpet flowers, while *Tabebuia serratifolia*, another tall tree, flowers when leafless and the whole crown is bedecked with long, bright yellow, foxglove-sized flowers. The long white flowers of *Bombacopsis quinata*, with petals and staminal tufts 10 centimetres long, are pollinated by bats and so open at night. The fallen petals and stamens of the previous night's flowers can be seen lying on the forest floor the next day. The buttercup tree (*Cochlospermum orinocense*) is another very conspicuous species. Like *Tabebuia*, it is deciduous and blossoms profusely, producing flowers like giant buttercups about 10–12 centimetres in diameter.

Anyone seeing tropical forest for the first time is impressed by the lianas, the great woody vines whose stems are suspended from the trees and often loop in bizarre fashion through the forest. Their sheer size, sometimes with stems 50 centimetres in diameter, astonishes the observer from temperate parts of the world. They belong to a number of families of which the Bignoniaceae (*Jacaranda* family) is the most important. Lianas are extremely floriferous and one often

The humming-bird pollinated flowers of the passion-flower, *Passiflora longiracemosa*. (*Gwilym Lewis*)

sees the canopy of the forest covered in thousands of large pink, mauve or yellow flowers of Bignoniaceae. There great climbers frequently have very large diameter water-conducting vessels in their stems, an adaptation to allow rapid flow to the leafy crowns through the immensely long and relatively narrow stems. This adaptation allows the forest-dweller to obtain pure water even when there are no streams or pools available. His technique is to hack off a length of liana stem, usually about 1 metre long, and simply pour water from one end of it directly into his mouth. Some species yield water much more readily than others, and the most productive is the *cipó de água* ('water liana'), *Pinzona coriacea* (Dilleniaceae), which is abundant in the Maracá forests. Aroids are also common climbers, and one species, a Swiss-cheese plant (*Monstera dubia*), is particularly conspicuous. As in some other climbers, the adult and juvenile foliage of *Monstera* are very different, and the casual observer would think they belonged to two different species. The young plant in the deep forest shade produces a slender stem with delicate, pale green, variegated leaves, closely appressed to the tree-trunk it is climbing. The adult foliage, on the other hand, is developed when more light reaches the plant, and consists of the well-known, large, leathery, perforated leaves.

Passiflora longiracemosa and a number of *Bauhinia* species are other noteworthy lianas of the Maracá forests. The former has long woody stems which produce their leaves only in the canopy but bear bunches of handsome, waxy, pink passion-flowers near ground level. Dr Forbes Benton studied the pollination system of this liana. By using chicken-wire cages of a mesh-size which allowed bees to pass but excluded humming-birds, he demonstrated that, (a) humming-birds were the only effective pollinators, and (b) the plants were self-incompatible (that is, although, like the majority of flowering plants, the flowers are hermaphrodite, the plant cannot be self-fertilized) (see page 104). The bauhinias are noticed by everybody visiting the Brazilian forests. They are very large leguminous lianas which have flattened, ribbon-like, regularly undulate, woody stems up to 50 centimetres thick and 100 metres in length. The leaves consist of twin lobes or leaflets, depending on how far they are divided, and because of this the name *Bauhinia* was given to commemorate the joint pioneer botanical work of the Swiss brothers Johann and Caspar Bauhin. The vernacular name of these lianas is *escada de jabotim* ('tortoise's ladder'), which always struck me as inappropriate since one could not possibly imagine a tortoise climbing one. However, William Milliken was recently told the following story explaining its origin by Waimiri Atroari Indians with whom he was living while conducting an ethnobotanical study:

> Mawa, the first Indian, decided to go and live in the sky. He summoned all of the animals to try to shoot an arrow high enough to pierce it. The only animal who succeeded was the *jabotim*. Mawa sent a woodpecker up with a *Bauhinia* vine to fix to the arrow. He then climbed up. As a gesture of gratitude he invited the *jabotim* to come up after him, but it was insufficiently agile to climb the vine. The spider-monkey, for whom the vine was no problem, started to climb. Mawa, who had not invited him, cut the vine when the monkey was half-way up and he fell with it into the tree-tops. They have remained there ever since [i.e. Mawa in the sky, the spider-monkey in the tree-tops, and the *jabotim* on the ground].

Lianas are opportunists which manage to place their foliage and flowers in the sun without expending the energy necessary to erect a free-standing trunk capable of supporting them. Epiphytes are plants which do the same thing simply by growing rooted to branches of tree-crowns rather than on the ground. However, the epiphytic habit has the disadvantage that water for the roots is scarce, and so the plants have adaptations which resemble those of desert plants (xerophytes). Epiphytes are most common in really humid forests, and Maracá, with its long dry season, is not ideal for them, so they are not abundant. Nevertheless, bromeliads (pineapple family), orchids and ferns are quite frequent as epiphytes in the trees.

Indians have a legend about the 'tortoise's ladder' liana (*Bauhinia* species) with its bizarre woody stems. (*James Ratter*)

An extreme development of the epiphytic habit is found in the stranglers. Here the life-cycle begins with the deposition of a seed by a visiting bird, usually in a crotch of a tree-crown. The seed germinates to produce an epiphytic shrub which sends down roots to the ground along the surface of the host tree's trunk. These roots branch and anastomose over the surface of the host's trunk, often forming a lattice-like woody network, and later by further growth and coalescence produce a complete enveloping woody cylinder. Where they reach the ground they usually form buttresses.

▶ Vivid flowers of *Allamanda nobilis*, a liana of the forest margin. (*William Milliken*)

▶▶ A strangling fig throttles its host tree, which will soon die. (*James Ratter*)

Tree-trunks have to grow by expansion to produce new tissues for transport of water and other substances, and the straitjacket of the strangler prevents the host from doing this. The host therefore dies and its tissues, now in the core of the enveloping trunk of the strangler, rot away while its crown disintegrates: the strangler has now usurped the place of its host in an act of extreme opportunism. The strangling habit is particularly characteristic of the fig genus (*Ficus*), and enormous buttressed strangling figs are common in the Maracá forests. However, it is also found in many parts of the tropics in such diverse and unrelated families as the Guttiferae, Bombacaceae and Myrtaceae. Clearly, this strategy for producing a tree by supplanting another has evolved independently in a number of families of the tropical forest. Sometimes the strangler establishes itself on an invulnerable host. Palm-trees do not produce new tissues by expansion of the trunk and are therefore very resistant to stranglers, so one sometimes sees a fig-tree with the fronds of a healthy palm emerging from the centre of its crown. The brick chimneys of old sugar-mills can also serve as supports for strangling figs, as can the pinnacles of abandoned temples in tropical Asia.

Lianas, epiphytes and stranglers can all be classed as mechanical parasites: they succeed in finding their place in the sun without the immense energy expenditure of producing a slow-growing, free-standing trunk. However, they are not parasites in the sense of taking nutrients from their hosts. True parasites in the form of a variety of mistletoes do occur in the Maracá forests, but are not as common there as in the savanna vegetation where small trees and shrubs act as hosts.

◀ Elaborate webbing of a tropical stinkhorn fungus, *Dictyophora duplicata*. (*William Milliken*)

◀◀ *Clathrus cancellatus*, a beautiful but evil-smelling fungus. (*James Ratter*)

◀ *Amaurodendron* fungus. (*Dudu Tresca*)

◀◀ Fungi such as this *Pycnoporus sanguineus* play a crucial role in extracting nutrient from litter on the forest floor. (*Duncan Scott*)

In general, the so-called lower plants are not conspicuous in the land environment. In places there are fair quantities of mosses growing on the bases of tree-trunks, while handsome toadstools are abundant on fallen trunks and branches, and large bracket fungi are found on dead and moribund trunks. The bracket fungi sometimes look like elephant's ears emerging horizontally from the tree-trunk and have the very appropriate vernacular name *orelha de pau*, meaning 'tree-ear'. Some of the fungi are luminescent, and at night their pale, eerie glow gives a fairy-like quality to the forest.

So far we have talked of the ordinary forest on drier ground, the *mata de terra firme* ('forest of firm land') which covers by far the majority of the island, but variations occur where there are differences in soils or topography and associated drainage.

Various species are characteristic of the riverine forest which lies along the two great arms of the Uraricoera River bounding Maracá Island, the Maracá and Santa Rosa channels. One has a wonderful view of this vegetation when travelling by boat. In areas where the river-bank shelves gradually there is a pure stand of *Inga ingoides* producing a shrubby thicket about 8 metres tall. The bushy shrubs making up the tangle do not develop erect trunks but are based on semi-prostrate boughs lying more or less on the surface of the muddy bank. During a large part of the year this thicket vegetation is partly submerged in the river. It is the home for noisy colonies of hoatzins (described in the next chapter) which crash about clumsily in the branches as the boat goes past.

Triplaris surinamensis, which to the casual eye re-

► *Coriolopsis* fungi attack dead wood. (*Dudu Tresca*)

►► Spines of the *tucumá* palm. (*Dudu Tresca*)

sembles a *Eucalyptus* in its pale bark and slender, erect form, is another common river-bank tree. It is the home of fierce stinging ants which protect it from all intruders and give it its vernacular name. The name is *pau de novato* ('novice's tree'), so-called because only novices make the mistake of trying to suspend their hammocks from it! *Triplaris* has male and female flowers on separate trees, and the female tree has flowers in great handsome pink inflorescences. It comes as a surprise to the non-botanist to learn that it is a relative of the humble temperate docks and sorrels. Cecropias also occur on some areas of the river-bank in habitats similar to *Triplaris*. They are also protected by ant-guards and, as well as occurring in the disturbed habitats of river-banks, are characteristic colonising species of areas where forest has been cleared for cultivation and later abandoned. The *Cecropia* tree is readily recognised by its candelabra form, jointed hollow stem filled with ants, and large palmately lobed leaves; it forms the principal food of the three-toed sloth which is obviously not discouraged by the ant-guard. Two species of river-bank trees are conspicuous by their great size and broad, deciduous crowns. These are the silk-cotton tree, *Ceiba pentandra*, which has already been mentioned, and a leguminous tree of similar form, *Enterolobium cylindrocarpum*. Both are also found in the *mata de terra firme* far from the river, but are probably somewhat commoner by the river-bank where the semi-open habitat gives room for their wide crowns. There is one final plant of this habitat which should be mentioned: a bizarre aroid called the *anhinga* (*Montrichardia arborescens*), with erect, rigid stems, reaching 3 metres, carrying large, leathery,

Costus arabicus, a relative of the gingers, often found in *buriti* palm groves. (*Dudu Tresca*)

triangular leaves, forms a narrow band along the water's edge.

In areas where swampy streams run through the forest there is a lower forest with a different floristic composition. The dominant tree of this community is usually the *buriti* palm (*Mauritia flexuosa*), a tall, graceful, fan palm often attaining 25 metres in height. The floors of these swampy areas are hummocked by superficial tree-roots and have abundant tufts of erect, dark, breathing roots of the *buriti*, resembling the bristles of inverted scrubbing-brushes emerging from the muddy soil. Such breathing roots are a characteristic of tropical forests on waterlogged soils and allow passage of oxygen to the submerged roots with which they are connected.

An almost treeless vegetation called *vazante* (see page 112) runs through the forest as strips up to 100 metres or more broad. It occurs along drainage lines so shallow no stream-bed has been cut, so that during the rains there is a slow-flowing marsh which dries up completely during the dry season. The vegetation consists of tall herbs covered in a tangle of herbaceous climbers. The whole aspect resembles an area of abandoned British allotments, covered in a rank mass of weedy vegetation, where decrepit sheds and disintegrating glasshouses would not seem out of place. There appear to be two components in the vegetation: semi-aquatic species which flourish during the rains and die back to their perennating underground parts in the dry season, and land plants which flourish in the drier periods. Notable among the former are the very tall, Marantaceous *Thalia geniculata*, reaching 3 metres, and a

Fungi caught in sunlight. (*Arno Lise*)

pale yellow-flowered species of *Canna*, *C. glauca*. Like so many aquatic plants, both of these species have wide distributions, ranging from the southern United States or, in the case of *C. glauca*, the West Indies, to northern Argentina. The dry phase of the *vazante* is dominated principally by tall grasses and tangles of herbaceous climbers belonging to the pea and morning glory families (Leguminosae and Convolvulaceae).

At Santa Rosa, at the east end of the island, there is a patch of true savanna vegetation, similar to that occurring on the neighbouring mainland. The area consists of grassland covered by a scattered layer of shrubs and small trees rarely more than 4 metres tall. These are of typical, fire-resistant, savanna type with characteristic contorted form and leathery leaves. They belong almost entirely to two extremely widespread species, *Curatella americana* and *Byrsonima crassifolia*, both of which stretch from Central America to Paraguay. *C. americana* has hard, silica-impregnated leaves which can be used as sandpaper (its vernacular name in Central Brazil is *lixeira*, which means sandpaper-bearer), while *B. crassifolia* is remarkable for its pleasant little yellow edible fruits. Both species have a considerable resistance to waterlogging, which is an unusual characteristic in Brazilian savanna trees, and indicates that during the wet season the water-table comes close to the surface of the sandy soil. The savanna area is crossed by a number of underground drainage courses followed by elegant lines of *buriti* palms. This type of landscape feature is known as a *vereda*. It is rare on Maracá Island, but enormous *veredas* with thousands of *buriti* palms can be seen on the mainland along the road between Boa Vista and the island.

◄ A flooded grove of *buriti* palms (*Mauritia flexuosa*), a familiar sight in the savanna near Maracá. (*John Hemming*)

◄◄ The view from the causeway leading to the ecological station: *vazante* wetland and a grove of palms at the forest boundary. (*John Hemming*)

The large *campo* area lying between the ecological station and the Uraricoera River is the first vegetation feature seen by visitors to the island. It is largely a mixture of *vazante* vegetation in the seasonally wetter areas and savanna vegetation in the drier.

Maracá Island is an area of contact between rainforest and savanna vegetation with, as one might expect, a climate transitional between the forest and savanna subtypes of the tropical rain climate. The forest of the island is fringe Amazonian rainforest right at its junction with the savanna vegetation which runs into Guyana and Venezuela. This is of great scientific interest as there are a fair number of studies of core Amazonian forest but very few of peripheral areas. Much has been learned of the history of the Amazonian forest during recent years. Whereas it was once thought that the forest had remained static for many millions of years, it is now known that it has been subject to dynamic changes. During the last glaciation in the temperate parts of the world, much drier conditions prevailed in what is now the wet tropics. This favoured the growth of savanna vegetation, and the Amazonian forest contracted to a number of isolated refugia occupying a small percentage of its present area. At the end of the glaciation, some 12,000 years ago, wetter conditions returned and the forest began to expand at the expense of the savanna, so that it now occupies its present enormous area of some 6 million square kilometres. The evidence on which this is based comes from a number of ingenious investigations in such fields as the study of the distribution of plants, birds, reptiles, recent fossil deposits, and even the geography of Indian languages. Work on the

► *Cattleya violacea*, an epiphytic orchid common on riverside trees. (*Steve Bowles*)

►► Fruits of a *Psychotria* species, a herb that grows in the shade of the forest floor. (*Dudu Tresca*)

► Forest fruits and seeds winged for dispersal by wind. (*William Milliken*)

►►► Biologists from southern Brazil studied grassy ponds on and near Maracá. Layers of vegetation in an almost dry wetland are: *buriti* palms (*Mauritia flexuosa*), cabbage-stalked *anhinga* (*Montrichardia arborescens*), and *Eichhornia crassipes* on the mud in the foreground. (*Haraldo Palo*)

southern fringe of the Amazonian forest in Mato Grosso demonstrates that the invasion of shrubby savanna (known as *cerrado* in most of Brazil) by the forest continues, when the situation is not masked by the influence of man. Similar studies are needed for the northern fringe and these can in part be provided by our observations on Maracá Island.

The reader will know of the great diversity of species in tropical floras, something first observed by the early explorers hundreds of years ago. We attempted to collect everything on Maracá, and our total was approximately 1050 species of flowering plants and fifty-two species of ferns and fern allies (see page 00). This included no fewer than fifty-four genera and 123 species of the family Leguminosae, intensively collected and studied by Gwilym Lewis of the Royal Botanic Gardens, Kew. The 1100 species of higher plants collected on the 110,000-hectare Maracá Island compares with approximately 1500 native species occurring in the whole of the British Isles. The Maracá forest is not as rich in terms of diversity per unit area as the core areas of Amazonia where up to 280 species of trees with trunks of 10 centimetres or more in diameter at breast-height are found per hectare. However, such forests are the most species-rich discovered anywhere in the world, and those of Maracá, with approximately eighty species per hectare, are fairly typical for more peripheral areas of Amazonia. Their diversity is, in fact, very similar to that of the forests on the southern fringe of Amazonia studied by the Royal Society/Royal Geographical Society Expedition to northern Mato Grosso more than twenty years ago.

Large numbers of the Maracá plant species are widespread in Amazonia, while others are more typical of the Guyanan flora. Some were recorded for the first time in Brazil, only a handful are completely new to science. This contrasts with some invertebrate groups described later in this chapter where novelties comprise a large proportion of the collections.

Much has been written on the abundance of alkaloids, tannins, terpenes and other chemicals in plants of the tropical rainforest. These distasteful and usually toxic substances function as a defence against herbivores, but are often also important to man as drugs (or spices). In fact the plants of the rainforest provide an enormous pharmacopoeia of which only a small part has yet been exploited (for example, quinine, curare). The rest remains to be investigated, but there is a great danger that it will be lost by habitat destruction before its value is realised. Here I shall mention just two of the plants with interesting alkaloids encountered during the Project. *Alexa canaracunensis* is an abundant small leguminous tree of the forests at the western end of the island. It contains the alkaloid castanospermine which has been shown to exhibit activity against the HIV virus *in vitro* and is therefore the focus of much interest from the medical world (see page 110). The second, *Ryania speciosa*, is a common small tree or large shrub of the forest understorey and belongs to a genus with almost legendary poisonous properties. I first came across *Ryania* in north Mato Grosso, where another species, *R. mansoana*, had a sinister reputation because the Indians were said to use it for poisoning each other. The poisonous principle, ryanodine, is so strong that the slightest amount ingested causes immediate death and therefore the plant's vernacular name is *mata calado* ('it kills silently'). According to backwoodsmen, throwing material of this plant on to a bonfire is sufficient to kill those who inhale the smoke, and it is certainly used to kill big caimans merely by rubbing the leaves on a meat bait. We were told of a tragic accident at a *churrasco* (barbecue) when a number of participants dropped dead immediately after sampling the steaks, and the problem was traced to the wooden spits which had unwittingly been cut from *mata calado*. As might be expected, the Brazilian army jungle survival course stresses the importance of avoiding this plant.

▲ White-lipped peccaries (*Tayassu pecari*) swimming to Maracá Island, where they are the most abundant large mammal. (*Fiona Watson*)

◀ The tapir (*Tapirus terrestris*), a gentle herbivore becoming endangered elsewhere in Brazil. (*Dudu Tresca*)

ANIMALS

DURING the Project about sixty zoologists worked on almost all the major animal groups of Maracá Island. The extensive results of their studies are being published in the report of the Ecological Survey of the island and in a host of articles in scientific journals. The following account gives the impressions of a non-zoologist. It is hoped that it will give some feeling for the habitat and the exciting animal life which fills it.

The sheer abundance of animals in the neotropical forests is enormous. This abundance, however, is not always obvious, except in the case of insects which everybody can see and feel. Casual visitors can crash through the forest for days in noisy, talkative groups without seeing any larger animals, since the latter are usually rather shy and the dense vegetation affords so much concealment for them. However, the skilled observer, staying still or moving quietly in the forest, sees many animals and can read the evidence of their presence in tracks, calls and other signs. The skill of the really good observer seems to be largely innate. It appears to be connected with an exceptionally independent temperament and, interestingly, a high proportion of the very best observers I have met have been left-handed.

The most common large mammal in the forest, and that most likely to be met by the visitor, is the white-lipped peccary, known locally as the *porco da mata* ('forest pig') or *queixada*. José Fragoso, a Portuguese mammalogist studying in the University of Florida, carried out a survey and discovered that the population density on Maracá was the highest ever recorded in any locality. Queixadas range the forest in aggressive, uncouth bands sometimes consisting of hundreds of individuals. These bands can usually be detected before they are seen by their powerful, pig-like odour and, once they have sensed one's presence, by the threatening gnashing of teeth. They are particularly common in the lower damper areas, often amongst the *buriti* palms, where they root about in the soil for both animal and vegetable food. There were a number of minor confrontations with queixadas during the expedition, and some people actually took to the trees to avoid them (see page 35).

Another, smaller species of peccary, the collared peccary or *caititu*, is also found on Maracá. It is altogether a much more attractive animal than its larger

◀◀ A rare glimpse of a brocket deer (*Mazama* species). (*Haraldo Palo*)

Jaguar (*Panthera onca*), Maracá's largest feline. (*Dudu Tresca*)

cousin, going about in small, perhaps family, groups that seem to be intent on minding their own business. The collar to which the name refers is a pale diagonal stripe which runs across the shoulder. Both species of peccary have enormous ranges, in the case of the *caititu* from the United States to North Argentina, and represent very successful, omnivorous non-specialists of the mammalian world.

Tapirs are by far the largest herbivores on Maracá and, judging by their unmistakable three-toed hoofprints, which one finds in wet places, must be quite common. Large patches of their horse-like, little-digested dung are also often found in the forest. Unlike the queixadas, however, they are quiet, solitary animals and are seldom seen. I only saw two during my visits: the first was basking in the shallows of the Uraricoera River, and we surprised the second in the forest so that it ran off in clumsy flight, crashing noisily into trees and other vegetation as it went.

Little brocket deer are also probably quite common in the forest. They are, however, shy, quiet, unobtrusive creatures, only about the size of roe deer, and are rarely seen.

We saw little evidence of large carnivores, but one day when I went to collect plants on the wet *campo* area close to the station, I was very surprised to find fresh jaguar dung within about 200 metres of the buildings. The director of the station, Sr Guttemberg Moreno de Oliveira, also had an encounter with a jaguar when fishing at night, while in 1989 a group consisting of one spotted and two black animals was seen swimming across the Maracá Channel.

Jaguars are powerful, spectacular beasts, but in general seem surprisingly unaggressive to man. Nearly twenty years ago, when our expedition was working in the forests of the Rio Suiá-Missu in southern Amazonia, jaguar tracks were everywhere, and we could hear the animals moving in the forest around us and sometimes calling. It was not uncommon to have the disconcerting experience of finding their tracks in the sandy soil on top of footsteps we had left only a few minutes before. Yet they were so good at concealment we never saw one, and their dogging of our tracks was probably a manifestation of cat-like curiosity. The only occasion I ever saw a jaguar was near dusk on a track in the Brazilian Pantanal when a fine spotted specimen bounded away from me like a big playful cat. Cynthia Sothers, a young botanist working on the World Wildlife Fund minimum conservation area study project based at Manaus, recently had an interesting encounter with one of these felines. She was returning to camp carrying an armful of plant specimens when she saw a jaguar on the trail ahead watching her with interest. Very naturally Cynthia stopped and she and the jaguar contemplated each other for a little time. She then put her specimens down and backed away while the jaguar came forward and sniffed and examined them. Cynthia took advantage of the beast's unexpected interest in botany to get off the scene and go to camp by another route. I do not know if she returned for the specimens later – somehow I doubt it!

The jaguars on Maracá presumably prey on the abundant *queixada* populations; however, queixadas are not easy prey since they live in large bands and show very aggressive group defence behaviour. The jaguars probably depend on pouncing on stragglers and must spend a lot of time watching for an opportunity. I was once shown a place where a jaguar had been observing a large band of feeding queixadas. It had concealed itself under a large bush and its tail, swishing back and forth in excitement, had brushed an area free of fallen leaves – exactly like a domestic cat watching birds in a garden.

One hears more tales of jaguars than of any other animal. I think most are true, and I could fill many pages with them. There are undoubtedly cases of people being killed and eaten by jaguars, but such cases are very rare and possibly the acts of exceptional rogue animals. However, they are a menace for anyone raising livestock, and this, together with the value of their pelts and the destruction of their habitat, makes their long-term future very insecure.

A baby spider monkey (*Ateles belzebuth*). (*Mark O'Shea*)

Andrea Nunes and Aléxia da Cunha recorded five species in their study of the primates of the island, while there is anecdotal evidence for the presence of one or two more (see page 62). The largest and most prominent primate is the red howler-monkey, or *guariba* as it is called in Tupi. Guaribas live in groups and produce their loud, eerie howling from the tree-tops principally at dawn and dusk, but sometimes at other times of day. The sound is produced by a bony resonating box in the throat and is quite unmistakable, resembling a very strong winter gale howling through the branches of trees; it was the characteristic early morning noise of the station. A pale-coloured species of spider-monkey is also common; it lives in small groups and is a remarkably bad-tempered creature which dislikes invasion of its territory by humans. When this occurs it leaps up and down on tree-boughs, like a furious chimpanzee, and sometimes throws sticks at the intruders. I was surprised at its marksmanship when one struck me with its first missile from a tree-crown some 15 metres away. Andrea Nunes made an intensive study of the behaviour of populations of these monkeys and obtained very valuable and interesting results. Capuchins (*macaco prego*) and the beautiful little squirrel-monkeys (*macaco de cheiro*) are also common in the forest. The night

► The primatologist Andrea Nunes observed squirrel monkeys (*saimiri sciureus*). (*Steve Bowles*)

►► Three-toed sloth (*Bradypus tridactylus*). (*William Milliken*)

Great ant-eaters (*Myrmecophaga tridactyla*) are often seen on the savanna near Maracá. (*Haraldo Palo*)

monkey is one of the species reputed to be present on the island, but was not observed by our primatologists. It is the only true nocturnal monkey in the world and has obvious adaptations to its way of life, such as large eyes. William Milliken and I were disturbed one night in a remote camp by a group of strange monkeys in the tree-tops; perhaps they were of this species.

A number of species of the ancient South American mammalian group, the Edentata, were found on the island. Great ant-eaters (*tamanduá bandeira*) are quite common, and, although they are more typically animals of open country, seem quite happy living in the forest. Their smaller cousins, the arboreal *tamanduás*, with long, slender, prehensile tails in place of the enormous bush of the terrestrial species, were also recorded. Both of these long-snouted ant- and termite-eaters have similar defence displays: they stand on their hind-limbs with their powerfully clawed fore-limbs spread wide apart ready to strike. They have no teeth to intimidate, but produce saliva which drips from their mouth during the display; the eyes are also closed, perhaps to protect them, or possibly for the same reason as the ostrich is said to bury its head in the sand. Anyone who disregards the threat will presumably come into contact with the powerful claws, designed to tear open hard termite nests, and the *tamanduá bandeira* also inflicts an almost unbreakable bear-hug which is strong enough to strangle a dog. The three-toed sloth is another Edentate which occurs on Maracá, although it seems to be rare. It feeds on foliage and spends its life hanging upside down high in the trees, usually motionless but occasionally

A three-toed sloth slowly descending from its favourite tree. (*William Milliken*)

moving very slowly. These characteristics, coupled with a good camouflage of dull tan fur tinted with resident blue-green algae, make the sloth almost invisible in the tree-crowns, hence the populations are perhaps far greater than we realised. Everything about the sloth is extremely slow and its Portuguese vernacular name of *preguiça* means the same thing as its English name. For a short time Mark O'Shea and Aléxia da Cunha kept one they had rescued in Boa Vista as a pet. In this role it failed miserably, although its owners were fond of it and defensively tried to point out intelligent behaviour. It could, of course, neither walk nor crawl but dragged itself laboriously along the ground. Most of the time, however, it was carried like a baby and waved its long, formidably clawed fore-limbs very slowly while contemplating the world with an inanely grinning, mask-like face.

Agoutis were the most commonly observed species of the very ancient and distinctive South American rodent fauna. They are slender-legged, buff-brown, almost tail-less animals with an arched back and the appearance of having the hind end of the body rather too big for the rest. They show some resemblance in appearance to the African duikers – perhaps an adaptive convergence of a rodent and a hoofed mammal in response to a similar forest habitat. Large agoutis weigh up to 4 kilograms or possibly a little more, and a large part of their diet is fallen fruits, eaten squirrel-like with the animal sitting up on its haunches and holding them in its front paws. Probably the most eerie sound in the forest is that of the howler-monkey, but the alarm cry of the agouti comes a close second. Never have I heard another sound which communicates such a sense of fear and foreboding. One of the legends of the backwoodsmen is that of the *curupira*, the mysterious demon of the forests, and the sudden hair-raising cries of an agouti pursued by a predator readily brings it to mind.

Adrian Barnett and Aléxia da Cunha studied the small mammals of the island, that is, those with a body weight of less than 5 kilograms, and discovered sixteen species. The great majority of these are rodents, mainly of small size, but including tree porcupines and a rare, large black furry squirrel. However, they also captured two marsupial species, an arboreal mouse opossum and a small species of terrestrial opossum, representing the most ancient and primitive mammalian group of South America (see page 106).

Fif Robinson and Rogério Gribel studied the bat fauna and discovered that it is extraordinarily rich (see page 106). No fewer than forty-eight species were recorded, ranging from species little larger than big moths, and weighing about the same as a British decimal penny coin, to spectacular leaf-nose bats of nearly 110 grams. Possibly this remarkable diversity is because the area represents the meeting-points of the Amazonian and Guyanan bat faunas. Fif and Rogério were extremely conscious of the importance of conservation. Their bats were captured by mist nets and non-damaging traps, and a wax impression technique invented by Fif was used to record dentition; after all the measurements and observations had been made, the animals were freed unharmed. These techniques contrasted with those of the Royal Society/Royal Geographical Society Mato Grosso Expedition of 1967–9 when the mode of collection consisted of attracting bats by waving a long, slender, flexible pole in the air and then firing dust-shot from a 16-bore Beretta into the swirling mass. We had been taught the pole-waving technique of attracting bats by Brazilian country people, and I imagine that it must function by some effect on the bats' echo-location system. One small species of bat, *Molossus molossus*, was very familiar to the members of the Project as it had taken up residence in great numbers in the roof-space of the buildings of the station. These bats were generally unpopular as their smell pervaded the air and their dried and crumbled faeces constantly rained through the joints in the ceiling. Neotropical bats are very variable in their ways of life, and at night, when most of us were sleeping (lightly covered with powdered bat faeces), the air was full of bats hunting insects, eating fruits, pollinating flowers, and even indulging in such exotic pastimes as catching fish or sucking blood.

The birds of Maracá have been studied by at least a dozen scientists, and the Brazilian zoologist Dr Debra Moskovits published a paper on the subject in 1985. During the Project about eight people worked on ornithology and a detailed report was written by José Maria Cardoso da Silva, an outstanding young ornithologist of the Emílio Goeldi Museum in Belém. José Maria recorded 444 species, at least one of these for the first time in Brazil, and it seems probable that the num-

A small leaf-nosed badger-faced bat. (*Duncan Scott*)

The Project's bat experts released their catches after careful measurement. (*Duncan Scott*)

ber might rise to as high as 600 if more intensive observations are carried out in the west of the island. Not only is the avifauna extremely rich, but great numbers of species can be seen in a short time – Dr Douglas Stoltz, of the University of Chicago and the Zoology Museum of the University of São Paulo, observed 219 species in a visit to the island of just three days. Thinking of this species diversity in comparative terms, the total number of species of birds in the world is about 8600, of which no fewer than 3500 are neotropical, while about 500 occur in the whole of Europe. So Maracá, with an area more or less twice the size of the Isle of Wight, probably has approximately the same number of species as the total European avifauna.

The birds of the terrestrial environment are so varied that only a few of the most conspicuous can be mentioned. If the *sororoca* ('wild banana') represents the vegetable marker of the Amazonian kingdom, then a bird called the *seringueiro* ('rubber-tapper'), or screaming piha, must be its animal equivalent. The *seringueiro* (*Lipaugus vociferans*) gives a loud, extended wolf-whistle, repeated at frequent intervals, which, more than any other, is the characteristic sound of the Amazonian forest. The bird itself is not easy to see and is relatively insignificant, resembling a long-tailed

▲ The presence of macaws shows that there has been little human interference at Maracá. (*Dudu Tresca*)

► A red-headed turkey vulture (*Cathartes aura*) – remarkably unaggressive for such a large and powerful bird. (*James Ratter*)

►► King vulture (*Sarcoramphus papa*), the largest Brazilian vulture. (*Haraldo Palo*)

◀ Laughing falcon (*Herpetotheres cachinnans*). (*Haraldo Palo*)

◀◀ Black-collared hawk (*Busarellus nigricollis*). (*Dudu Tresca*)

female blackbird. Parrots are well represented, and include three species of enormous macaws which fly over the forest, usually in pairs, uttering their raucous cries of 'arara', which is also their onomatopoeic Brazilian vernacular name. One of the species is the common blue and yellow macaw, which is very widespread in Brazil, while the others are predominantly red species. The blue and yellow macaw nests inside the top of dead *buriti* palm-trunks, a situation well protected from ground predators.

The widespread red-headed turkey vulture and its two very similar congeners, the greater and lesser yellow-headed vultures, spend their days drifting about in the thermals awaiting the appearance of carrion, while the less common king vulture (*urubu rei*) does the same thing but flies at a much greater height. The king vulture is a very large, spectacular, black and white bird with a garish multi-coloured head and resembles a reduced version of its giant relative, the condor. It has no sense of smell and depends on sight to find the carrion on which it feeds. This would be a great disadvantage in forest country if it were not that the king vulture watches, almost literally with eagle-eye, the smaller turkey and yellow-headed vultures below. These smaller vultures have a keen sense of smell and when they detect carrion and descend, the king vulture follows them and, because of its greater size, is able to chase them off their food. Thus one finds king vultures feeding on carrion below a dense tree-canopy, in places where they could not possibly have found it themselves without the unwitting aid of their smaller cousins.

The *anu preto*, called the smooth-billed ani in English,

is probably one of the first two birds noticed by the casual visitor to Brazil, the other being the black vulture. It is a fair-sized, long-tailed, jet-black cuckoo, with a crested bill giving a rather hornbill-like appearance. *Anu pretos* are great survivors in a changing environment and have adapted to man-made habitats, particularly scrubby abandoned cultivations. Their natural habitat seems to be areas of low vegetation in dampish places, such as *vazante*. They always live in small flocks, often of six to ten or more birds, which typically can be seen busily flying and gliding from bush to bush of scrubby areas. They often associate with cattle, and their generic name is *Crotophaga* (Greek, 'tick-eater'), although doubt has been cast on whether they are particularly active in picking ticks from the backs of the cattle on which they perch. *Anu pretos* are communal nesters and the females of the flock usually lay in a big, deep nest, which can contain up to fifty eggs, although ten to fifteen is more usual. Apparently, however, each female incubates her own eggs, and there seems to be some use of leaves to partition the clutches. The eggs themselves are surprisingly large, up to 20 per cent of the body-weight of the female. The communal nesting habit found in many cuckoos is a precursor of the parasitic habit – presumably it is only going one stage further to deposit the eggs in someone else's nest.

Fifteen species of humming-bird were recorded from the island, but although they are easy to see in more open vegetation such as savanna and *vazante*, they are surprisingly difficult to spot in the forest. Often one only hears sudden, rapid bursts of their wings close at hand without succeeding in seeing the bird. This unexpected noise in the forest can seem menacing to the tiro who does not know what it is. One sometimes sees the beautiful nests of a species of humming-bird in savanna shrubs; they are neatly made, little bigger than thimbles and are camouflaged with lichens which are incorporated into their outer layers. The Brazilian vernacular name for humming-bird is *beija flor* ('it kisses the flower'), an extremely apt title when one sees the birds going from flower to flower, giving each a touch with the beak.

Mark O'Shea carried out an energetic herpetological survey of the island, and incorporated the fine work already done by Márcio Martins, Celso Morato and Debra Moskovits (are all herpetologists' names so rich in Ms?). Incidentally, Mark fits the image of an intrepid explorer far more than any of the other British participants in the Project, although hardly to the extent of the Brazilian Victor Py-Daniel (see next chapter). He is a wiry, thin-faced, freckled young man with a long, luxuriant red beard, who wears an Anzac hat and is festooned with exotic derring-do equipment such as crossbows and red-bladed machetes. His enthusiasm for his subject and ability to catch and handle poisonous snakes are truly phenomenal (see pages 29 and 104).

The species of reptile recorded consisted of thirty-four snakes (including at least one new species), twenty-seven lizards (one new), two tortoises, five turtles, two crocodilians and three of the worm-like burrowing amphisbaenans. The amphibia consisted of nearly forty species of frogs and toads. Many of these animals are primarily aquatic and so are discussed in the next chapter.

Apart from scuttling lizards, the most obvious reptiles on Maracá are the two species of big tortoises (*jabotim*). They are abundant in the forest, grow to a length of over 40 centimetres, and, as can be imagined, attain a considerable weight. A number of years ago Dr Debra Moskovits spent a year on Maracá carrying out detailed fieldwork on the tortoise populations, and many of the wild tortoises still carry her painted identification numbers and notches on their shells. These tortoises are a popular food of the local Indians: when travelling in their canoes they carry live tortoises which are slaughtered at mealtimes. The tortoise thus provides fresh meat in a convenient container.

Large rattlesnakes, up to about 1.5 metres in length, are quite common in the forest near the station, and a number were temporarily captured and studied. During its captivity one bit Mark O'Shea on the hand, but injections of antiserum by our nurse, Sarah Latham, saved him from any serious effects, as described in the previous chapter. These big rattlesnakes have poison fangs up to 3 centimetres long and inject a large quantity of venom, so that anyone would be very lucky to survive their bite without treatment with antiserum. Another highly poisonous snake, the *jararaca* (*Bothrops atrox*, usually called the Fer-de-Lance or Tommy Goff in English), was observed on several occasions and collected once. With the exception of the small coral-snakes, all poisonous South American snakes (rattle-

◄ A Red-footed tortoise, retracted into its shell. (*William Milliken*)

◄◄ red-footed tortoise (*Geochelone carbonaria*). (*Haraldo Palo*)

◄ Three Mount Roraima rattlesnakes (*Crotalus durissus*). (*John Hemming*)

► Fangs of the rattlesnake that bit Mark O'Shea. (*James Ratter*)

►► A rare and secretive mop-headed iguanid (*Uranoscodon superciliosum*). (*Haraldo Palo*)

snakes, bushmasters and the various species of *Bothrops*) are pit-vipers, so-called because they have a heat-sensitive pit situated on each side of the head between the nostril and the eye. These pits are used for detecting warm-blooded animals, which form an important part of their prey.

I am not one who appreciates the beauty of snakes, but must admit that the patterning of the non-poisonous boa constrictor seen in the forest is most attractive. It has a series of large, more or less chocolate-coloured saddles with light centres along its back, and a series of similar smaller blotches along its sides, and terminates in a light reddish tail. Boas feed on small mammals, birds and lizards and can grow to a massive 4 or 5 metres in length (smaller, however, than their cousins, the anacondas, discussed in the next chapter). The single specimen captured on the Project showed evidence of an encounter with awkward prey: it had five porcupine quills buried in the roof of its mouth, one protruding through the top of its head.

Many of the frogs encountered were little, dainty, agile, hylid tree-frogs, camouflaged to blend into the forest background, but a spectacular little arrow-poison frog was also found. *Dendrobates leucomelas* advertises its toxicity by striking black and yellow warning colours; it is one of a group of frogs which Indians use to prepare arrow-poisons for hunting. A rather contrasting amphibian, the far from dainty or agile great marine toad, is common on Maracá, as it is in most parts of Brazil I have visited. It is a squat, warty, somewhat unattractive animal of rather unremarkable toad-colour, reaching about 22 centimetres long and perhaps 16 centimetres

◀ Boa constrictor (*Corallus enydris enydris*). (*Fiona Watson*)

◀◀ A green Lora snake (*Philodryas olfersii*). (*Mark O'Shea*)

▼A brilliantly coloured Brazilian rainbow boa (*Epicrates cenchria*) collected close to the station. (*Mark O'Shea*)

► A tree frog. (*William Milliken*)

►► Tree frog (*Hyla crepitans*). (*Mark O'Shea*)

► Marine toads (*Bufo marinus*) are common throughout Maracá Island. (*Haraldo Palo*)

►► The tiny *Dendrobates leucomelas* is known as the arrow-poison frog because Indians use its powerful poison as an ingredient in curare.(*William Milliken*)

► A Hylid tree frog (*Hyla boesemani*). (*William Milliken*)

broad. It hunts at night and loves to hop around under electric lights gobbling up the hundreds of disoriented and singed insects to be found there, moving about very confidently as the venom secreted from the skin-glands more or less exempts it from attack by predators.

Nobody could fail to be impressed by the insects of the neotropics. They are omnipresent and abound in myriads of forms, from beautiful metallic blue *Morpho* butterflies and giant rhinoceros beetles to jigger-fleas, which bury themselves in one's toes, and tiny powder-like biting sand-flies. The colonial species build huge earthworks or arboreal nests and carry out Herculean labours such as consuming dead tree-trunks or harvesting tons of leaves from the forest. They transmit a host of parasites, some of which are responsible for appalling diseases. Careful examination demonstrates adaptation at all levels: sombre insects show amazing camouflage, while others parade in garish warning colours, either because they are genuinely dangerous or distasteful, or because they gain protection by mimicking a model which has these properties. Incidentally, it was observation of the widespread occurrence of the latter phenomenon in Amazonian insects that caused the great nineteenth-century naturalist H.W. Bates* to formulate his theory of mimicry.

About 900,000 species of insect have already been described, and estimates of the total number are continually being revised upwards to perhaps even as high as 30 million. In Amazonian rainforest insects represent almost 50 per cent of total animal biomass, and since their individual weight is usually infinitesimal in comparison to higher animals, their sheer numbers are absolutely astronomical. To accommodate this diversity, more than half the zoologists on the Project worked on insects, and they made huge collections which are at present being studied. So far about 1700 species have been identified, of which perhaps about 8 per cent are new to science. Obviously we can only mention a few selected groups here.

One of the most prominent groups of insects in the tropics is the order Hymenoptera which includes the ants, bees and wasps. Ants really dominate many aspects of life in the tropical forest. The huge underground nests of the leaf-cutting *saúva* ants, which must be the home for millions of individuals, are the centre from which working parties foray out each night to collect leaves. The forest floor is covered by their well-swept trails which can sometimes be mistaken for paths by walkers. Most of the leaf-cutters' work is done during the hours of darkness, but early in the morning columns of workers can still be seen returning to the nest, carrying little cut pieces of leaves or sometimes flowers over their backs. Great quantities of these cut leaves are jettisoned on the path, presumably by ants which decide that they have run out of time and must return to the nest with the greatest possible speed. The

A large palm-boring beetle (*Rhynchophorus palmarum*). (*Haraldo Palo*)

Locust (Acrididae family). (*Haraldo Palo*)

A rhinoceros beetle. There are vast numbers of beetle species in rain forests. (*William Milliken*)

* When Henry Walter Bates returned from his eleven years' collecting in Brazilian Amazonia, he became the first paid Secretary of the Royal Geographical Society. He served the RGS in this way for some thirty years. John Hemming is only the fifth incumbent of this post created 129 years ago.

► A long-horned grasshopper (Tettigoniidae family) perfectly mimics a leaf and twig. (*William Milliken*)

►► Damsel-flies mating. Prolific insects far outweigh mammals in biomass in tropical forests. (*William Milliken*)

► Leaf-cutting *saúva* ants (*Atta* species) carrying sail-like pieces of leaf. (*Dudu Tresca*)

►► Army ants (*Eciton* species) kill anything that does not flee from their column. (*Steve Bowles*)

leaf-cutter ants use the leaves as a medium to grow the fungi on which they feed, and their nests contain galleries of fungus gardens.

Two species of army ant (*Eciton*) are common on Maracá, swarming over the forest floor and killing any animal in their path which cannot make its escape and is small enough for them to handle. They are not very large, but are ferocious, particularly the soldiers which have long, powerful hooked mandibles, so that a few *correição* getting into one's trousers is something to be remembered. They do not produce any permanent nests but merely mass their bodies together in large resting bivouacs. Another disastrous ant to tangle with is the *taióca*, a large species 2 centimetres long which travels around the forest floor in comparatively small foraging parties of about twenty individuals, and can both bite and sting.

There are many tiny species of colonial bees belonging to the genera *Trigona* and *Tetragona* which usually live in nests in hollows of trees. Although some species are no bigger than a fruit-fly, they are perfect bees in miniature, complete with sometimes disproportionately large pollen-baskets on their back legs. They are attracted to perspiration and often swarm over people working in the field, lapping up the sweat with their long tongues and thus producing a slight tickling sensation. Sometimes the numbers trying to enter one's nose, ears and even eyes make work in the field extremely uncomfortable. They have no stings, but for such tiny animals have a surprisingly strong bite which they use when defending their nests. They also have a distinctive alarm odour, reminiscent of the smell of cattle-cake; this is produced, for instance, when one is accidentally crushed. The Brazilian vernacular name for these little creatures is *lambe olho* ('it licks the eye').

Fortunately we did not encounter the great colonies of the fierce, introduced African honey-bees near the station, although I was told that some occur towards the west of the island. I had a lot of experience of these aggressive insects in northern Mato Grosso twenty years ago, and they were undoubtedly by far the most dangerous animals in forests which abounded in wildlife, including many jaguars. The worst stingers in the Maracá are various species of wasps and hornets, but these are not aggressive unless they feel their nests are threatened.

The blue *Morpho* butterfly is a beautiful sight in the shade of tropical forests. (*James Ratter*)

A cryptic moth is camouflaged as leaf litter, visible only by its antennae. (*William Milliken*)

The eye marking on an owl butterfly's wing (*Caligo eurilochus*) is intended to confuse predators. (*William Milliken*)

► A caterpillar of the family *Limacodidae* mimicking insect eggs. (*James Ratter*)

►► The bright colouring of a hairy orange caterpillar is probably a warning to predators that it is poisonous. (*Mark Rickards*)

► A butterfly chrysalis hanging from a leaf. (*Arno Lise*)

►►A peanut bug (*Lanternaria species*). (*Gwilym Lewis*)

►►► Hawk moth caterpillar (Sphingidae family). (*James Ratter*)

►►► Tarantula spider. (*William Milliken*)

The butterflies are another conspicuous group showing enormous diversity of species and abundance in numbers. Once or twice in Brazil I have witnessed migrations when the air was full of butterflies for 10 or 20 miles, and the sheer number of individuals was incalculable. Drs Olaf Mielke and Mirna Casagrande of the Federal University of Paraná collected no fewer than 2724 specimens on Maracá belonging to 458 species, of which at least one is new. This is a very low level of novelty compared to other insect groups. It is explained by the fact that butterflies have already been more intensively studied than most other neotropical insects since they are conspicuous and, like birds, attract human attention. The largest and most spectacular butterflies on Maracá are the metallic-blue morphos and the owl-butterfly (*Caligo eurilochus*), so-called because of two large eye-spots on its hind-wings.

There are hosts of other interesting insects, including the ecologically important termites, giant Cerambycid beetles 8 centimetres long with great thick antennae longer than their bodies, big Peanut bugs (*Lanternaria* sp.) with devices to frighten would-be predators in the form of a false reptilian head at one end and large eye-spots at the other, and wonderful mole-crickets which try to squirm out of one's grip by vigorous rowing movements of the spade-like front legs, but neither space nor the scope of this account makes it appropriate to attempt to describe them all.

The spiders of the island range from hand-sized, hairy tarantulas to numerous minute, inconspicuous species. Professor Arno Lise of the Zoobotanical Institute of Rio Grande do Sul collected about 275 species of which a high percentage are new to science (see page 102).

The last animal to be mentioned is also an arachnid, and the curse of fieldwork in Brazil. During the dry season, ticks (*carrapatos* in Portuguese) swarm in the Maracá forests, and after a day in the field people usually remove anything up to a few hundred of them from their bodies. One learns to detect the presence of ticks almost instinctively and usually they are removed before they can bite. If they do succeed in biting, it takes them a number of days to fill with blood until they look like little grapes. In humans this only occurs if the tick manages to find a very concealed place, such as in the hair. The greatest nuisance is caused when a person

Spiders' webs come in many curious shapes. (*Dudu Tresca*)

The beautiful symmetry of a spider's web. (*Arno Lise*)

◄◄◄ Professor Arno Lise discovered twenty-six new species of spider on Maracá. (*Arno Lise*)

◄◄ A Turbellarian flat-worm. (*Arno Lise*)

◄ A tick bloated with human blood. (*James Ratter*)

encounters larval ticks just hatched from the eggs. Generally when this happens an enormous number is picked up, often representing all the offspring of an adult female which died after laying its eggs. The tiny larvae are called *mucuim* or *carrapato-pólvora* (the latter meaning gunpowder-tick) and are only just visible to the naked eye. They have only six legs in contrast to the eight of the later stages in the life-cycle – perhaps there is no room on the tiny body to fit on another pair! Removing mucuinhos from the skin and clothing when tired after a long day in the field is a very tedious chore.

So far nothing has been written on personal feelings towards the forest. Obviously this is a very idiosyncratic subject which will vary from individual to individual, but I think some general observations can be made. Recently I read a book on the Amazonian forest in which the writer stressed that man always feels an intruder in a somewhat hostile environment. I have always found the opposite: something that constantly surprises me is how quickly the area of forest where one is working becomes home, and how one regards an initially strange and apparently featureless patch of forest as an interesting and friendly garden after spending a couple of days in it. One feels that the forest with its lack of traffic and man-made bustle should produce a great feeling of tranquillity, but this, alas, is seldom the case for the botany survey teams, who always seem to be working against the clock to fit their work into the allotted time! Nevertheless, on days off one can sometimes enjoy this tranquillity – on one such day I went by boat on a picnic trip along the Maracá Channel and found the myriad of small forested islands one of the most magnificent sights of my life. Those who live and work in the forests have a great respect and affection for them: José Ramos, the INPA *técnico*, put this into words on one occasion by saying that after twenty years of travel, the place where he felt most at home and happiest was in the tall *terra firme* forest.

CHAPTER FOUR

The Aquatic Realm

WHEN writing about Amazonia everything is expressed in superlatives, and never is this more true than when describing the aquatic environment. The Amazon basin is some 7 million square kilometres in area and is drained by the largest river in the world. The main course of the Amazon–Solimões* is 6500 kilometres, some 200 kilometres shorter than the Nile, but in terms of water flow it is about four times greater. In fact at least 20 per cent of total freshwater discharged into the oceans comes from the Amazon. Not only is the Amazon–Solimões the greatest river in the world, but two of its tributaries, the Rio Negro to the north and the Rio Madeira to the south, are the second and third largest. Because the pattern of rainfall is seasonal and the relief remarkably flat, the Amazon basin is subject to an annual cycle of floods and low-water periods. Perhaps one of the best ways to appreciate the immensity of Amazonia is to fly from Belém to Manaus on one of the rare clear days at the height of the floods in February or March and see the vast watery kingdom below with its huge areas of flooded forest and *campo*.

The rivers of the Amazon basin are classified into three categories according to the types of sediment they carry, and this in turn is determined by the terrain through which they flow. White-water rivers have muddy sediment, while black-water rivers have the characteristic dark tint of tannins from decomposing vegetation and clear-water rivers have little suspended sediment. The white water contains the highest nutrient levels and hence the richest aquatic life, while the black waters are poorest. Where large rivers of different colours meet, the waters often run a considerable distance before mixing. The best-known example of this is at the meeting of the Rio Negro (black water) and the Solimões (white water) close to Manaus.

As we have seen, Maracá Island is a 110,000-hectare fluvial island formed by the bifurcation of the Uraricoera River into two channels, the northern Santa Rosa Channel and the southern Maracá Channel, which reunite 60 kilometres downstream. The Uraricoera is a clear-water river, lacking the dark peaty material found in many other affluents of the Rio Negro. It is formed by the joining of the Rios Parima and Aracasa, which originate in the Serra do Parima, and joins far downriver from Maracá with the Rio Takutu to form the Rio Branco which in turn flows into the Rio Negro. The course of the Uraricoera River from the union of the Parima and Aracasa to the junction with the Takutu is approximately 400 kilometres – a remarkable length when one remembers that it is only an affluent of an affluent of an affluent of the main Amazon!

The Uraricoera River has innumerable rapids, rocks and secondary channels, causing great problems for river travellers. During the year of the Project, as previously mentioned, many of the team became expert in

◀ The rocky bed of the Maracá Channel exposed during the dry season. (*Haraldo Palo*)

* The main course of the river is called the Solimões above its junction with the Rio Negro.

► Shooting the Tipurema rapids. Passengers and cargo have been unloaded. (*Dudu Tresca*)

▼ The many rapids surrounding the island helped keep Maracá uninhabited. (*Chris Caldicott*)

navigating the difficult river, having learned the correct courses through rapids and other obstacles from Maiongong Indians who worked with us from time to time. Luis Pestana, who had once been a professional sailor on the Brazilian north-eastern coast, Amazonas, and the Macuxi Indian Xico were accomplished helmsmen, while William Milliken developed a remarkable knowledge of the rivers. William recorded the routes used by the Maiongong and other experienced helmsmen in a guidebook for river-travellers. This book maps the entire recommended course circumnavigating Maracá Island, with details of twenty-one rapids and other obstacles on the Santa Rosa Channel and no fewer than 104 on the Maracá Channel. River knowledge allowed a great reduction in journey times, so that towards the end of the Project William and Amazonas succeeded in travelling from one end of the island to the

◀ A calm stretch of the often turbulent Maracá Channel. (*Dudu Tresca*)

Sunlit riverine vegetation. (*Dudu Tresca*)

other in one day – a journey which would have taken three or four days a year earlier.

Navigation of the northern Santa Rosa Channel is easier than that of the southern branch, the Maracá Channel. Over much of its length it is broad – up to 100 metres or more – and calm, but at intervals there are rapids through which water rages and boils around granitic rocks. Such rapids are called *cachoeiras* in Portuguese, and on the Uraricoera they have romantic-sounding names such as Tipurema, Português, Apertada na hora ('suddenly narrowed') and Fumaça ('smoke'). Many of the rocks occurring in the *cachoeiras* and elsewhere in the river are giant, rounded granite boulders showing characteristic 'onion-skin' weathering (see the photograph on page 81). When the water was high during the rainy season, the passengers usually disembarked and took a land-route past the *cachoeiras*,

Robert Schomburgk described the Purumame falls as one of the finest in the Guianas. (*John Hemming*)

while the long aluminium boat, Mercury outboard roaring, raced in a cloud of spray through the boiling waters under the skilled control of the helmsman and the guiding *proeiro* (prowman). After completing the passage the helmsman traditionally took a slug from a *cachaça* bottle – an example usually followed by the rest of the crew who, considerately, did not want the helmsman to feel guilty at drinking alone. Thus the passage of a series of *cachoeiras* was quite a cheerful business, and fortunately there were no breathalysers on the Uraricoera. In the dry season, the passage of the *cachoeiras* could be the other extreme, with the boats having to be dragged and carried over the rocks: an exhausting exercise at temperatures of 40°C (over 100°F)!

The Maracá Channel also has broad calm stretches and spectacular *cachoeiras*, but for much of its length the most striking feature is the disintegration of the main course into a labyrinth of narrow channels where water races between forested islets. Nobody but Maiongong Indians or intrepid explorers armed with William Milliken's maps could find their way through this maze.

I never visited the western end of the island so did not see the most imposing feature on the Uraricoera, the legendary Purumame Falls (see the photograph above and descriptions on pages 7 and 26). William Milliken describes them thus:

> At the western extremity of the island lies a narrow range of abrupt hills, densely clothed in a forest that holds little in common with the forests of lowland Maracá. The Furo [Channel] de Santa Rosa, shortly below the point of its divergence from the Furo de Maracá, tumbles through a narrow gap in the range in three spectacular foaming leaps, at the base of which lies an idyllic sandy beach.

Approached from below, these falls afford an impressive and unexpected sight as one rounds the last bend in the river.

Aquatic plants of the Podostemaceae family cling desperately to the near-sheer rocks below the rushing current, and a luxuriance of ferns enjoys the clouds of spray thrown up along the banks. The forest looms precariously above, casting much into shadow, but sunlight illuminates the spume and a roaring thunder fills the gorge. Below, in the deep swirling pool at the base of the falls, was caught a monstrous black piranha with a mouth little smaller than that of a human, and teeth like razors.

River travel was probably the most romantic and certainly the most explorer-like activity of the Project. Some excelled at it and, had they been less modest, could have told ripping yarns of capsizing and other hairy incidents in the *cachoeiras*. None of the crews was more spectacular than that of Dr Victor Py-Daniel, the medical entomologist of INPA, who made long voyages on the Uraricoera River and its tributary the Uraricáa to study the simulid flies which carry river-blindness. Victor, a jovial, bearded figure in paramilitary uniform, 6 feet 4 inches tall and weighing over 15 stone, came close in appearance to caricatures of a colonel of mercenaries or perhaps a late-twentieth-century Professor Challenger.

The beauty of the river can be quite breathtaking, whether on the long broad stretches of the Santa Rosa Channel, with low hills clothed in tall forest along the banks, or wending one's way through the intricate labyrinths of the Maracá Channel, with forest canopy overhead. The vegetation has already been mentioned in the previous chapter, but one's attention is sharply focused on it when battling along these tunnel-like mill-races through the forest. Tall, slender *buriti* (*Mauritia flexuosa*) and *açaí* (*Euterpe precatoria*) palms are conspicuous, as is the leguminous tree *Elizabetha coccinea*, whose scarlet and white flowers and large marzipan-coloured pods stand out from the surrounding green, and the abundant *Coussapoa villosa* with branches just above the water's surface bearing conspicuous rhomboidal leaves. Lianas are abundant in the overhanging vegetation and include the spectacular crimson flower-spikes of the *escova de macaco* ('monkey's brush', *Combretum rotundifolium*, see the photograph on the right), *rabo de arara* ('macaw's

A tangle of lianas in the riverside vegetation. (*John Hemming*)

Spectacular scarlet inflorescences of 'monkey brush' liana (*Combretum rotundifolium*) were often seen overhanging the river. (*Dudu Tresca*)

Conspicuous scarlet extrafloral nectaries of the 'macaw's tail' liana (*Norantea guianensis*). (*William Milliken*)

► Leaf of an *anhinga* (*Montrichardia arborescens*). (*Dudu Tresca*)

►► A water hyacinth (*Eichhornia azurea*). (*Haraldo Palo*)

► Veins and necrotic patches on an *anhinga* leaf. (*William Milliken*)

tail', *Norantea guianensis*, page 85) with 60-centimetre flower-spikes covered in vermilion slipper-shaped nectaries, a woody *Machaerium* species with viciously hooked spines, and the leguminous *Entada polystachya* festooning its host trees with big straw-coloured pods 30 centimetres long. In places where the water flows more slowly the *anhinga* (*Montrichardia arborescens*), lines the water's edge (see above and page 52). The water (*Eichhornia crassipes*), a floating plant with swollen leaf-stalks and clusters of handsome pale blue flowers, is often found in rafts lodged amidst the waterside vegetation. This species has been introduced to many other parts of the tropics where it has multiplied without check to create serious problems by choking rivers, reservoirs and other stretches of water. In Amazonia, however, it is native and in harmony with the environment.

Species of the family Podostemaceae are found in the inhospitable habitat of the rapids. These remarkable submerged plants look rather like seaweeds firmly attached to the rocks where the flow is fastest. Their inflorescences are borne above the water, and the largest species, *Mourera fluviatilis*, has 30-centimetre spikes of pink flowers which stand trembling above the fast-moving water. They are extremely attractive to large bees which fly like bullets right out to the middle of the river to visit them.

In addition to flowering plants, many species of algae occur in the rivers and other aquatic habitats. Drs Pedro Mera and A. Oliveira of INPA studied them as part of the Project and found no fewer than 232 species.

The diversity and abundance of animal life in the Uraricoera is enormous, showing the rich neotropical fauna in its natural state without the influence of hunting or other exploitation.

The pink river dolphin (*boto*, *Inia geoffrensis*) is amongst the largest predators of the river. It is a pale pinkish white animal, growing to over 2 metres in length, with a long beak and a more distinct neck than marine dolphins. The dorsal fin is a mere ridge, rather similar to that of a sperm whale. *Botos*, like marine dolphins, are interested in people and often approach to investigate bathers. This seems to have led to the story that they will help a swimmer in distress, which is possibly true, and another legend which most certainly is not. According to the second tale, male *botos* transform themselves into handsome young men who participate in the river-village *festas* and are irresistible to the girls. Their visits often result in the birth of children who show no sign of their miscegenation. The fathers, of course, cannot be traced, as when it comes to holding the baby they are cavorting in the river chasing fishes! Such children are called *filhos de botos* ('children of dolphins') and no attempt is made to determine their paternity more closely – a useful device for avoiding social tensions in small communities. According to legend, one way to recognise a transformed *boto* is that he always wears a hat to hide the blow-hole in the top of his head. However, it seems strange that the *boto*, having shown the ability to change into human form, acquire elegant clothing and presumably indulge in fluent speech, could not alter this last little detail. Some country people actually believe this astonishing superstition, and unfortunately it has caused the destruction of some *botos*, animals which are otherwise generally respected and left alone. The killers are not outraged fathers but young men who think that having a *boto*'s eye in their pocket will endow them with the animal's irresistible charms.

A smaller species of the dolphin, the *tucuxi* (*Sotalia brasiliensis*) is also found in the Amazon system, as well as in the sea along the Atlantic coast of South America. It is a darker animal than the *boto*, with a well-developed dorsal fin, and is the world's smallest cetacean. It is commoner than the *boto* in many parts of Amazonia but is rather rare in the Uraricoera River.

Two species of otter also occur in the Uraricoera, one of which is the giant river otter (*ariranha*, *Pteronura brasiliensis*). *Ariranhas* are shy, slender, almost seal-like animals, reaching nearly 2 metres in length. They have been hunted to the verge of extinction for their valuable skins but they still occur around Maracá Island, although they are rare. Dr Elton Colares of INPA studied *ariranhas* as part of the Maracá Project and obtained results contributing to the rather sparse existing knowledge of the animal.

Elton Pinto Colares from INPA studied the magnificent but endangered giant otters (*Pteronura brasiliensis*). (*Dudu Tresca*)

Capybaras (*Hydrochaeris hydrochaeris*), the world's largest rodents, are quite common on Maracá's river-banks. (*Dudu Tresca*)

Other mammals typically associated with the river are tapirs, which spend a great deal of their time in water, and capybaras, the world's largest rodents (more or less the size of small pigs), which are even more highly aquatic. In addition, tiny bats, little larger than moths, roost under rocks and boughs overhanging the water and are disturbed when boats are passing through narrows. A more delightful sight than a squadron, or at least flights, of these minuscule mammals flitting along the channel in front of a boat can hardly be imagined. I had read of these little animals in Peter Fleming's *Brazilian Adventure* nearly forty years before I ever saw

them. Fleming was also deeply impressed by them. He wrote:

> They were the smallest and flimsiest creatures imaginable, so grey, so silent, so altogether negative, that you could hardly be certain that you had seen them at all. . . . In the air they keep close together, flitting urgently through the alien sunlight in a little grey cloud with no more substance to it than a puff of vapour.

More scientifically, they are called *Rhynconycteris naso*, their wingspan is about 12 centimetres and their weight approximately 4 grams (about the same as a British decimal penny). *Noctilio leporinus*, the fishing bat, also occurs on the Uraricoera. It is quite a large bat, some twenty times heavier than the previous species, and can be seen at dusk flying low over the water, every now and then touching the surface to gaff small fishes with its oversized clawed feet.

The wide vistas of the Uraricoera are wonderful places to observe birds. One often sees absolute aviaries of them when sailing along the river: not only aquatic species but also terrestrial birds crossing from bank to bank or, in the case of vultures and storks, soaring in the thermals. This contrasts with the closed forest where vegetation limits the view to a few metres and the sky is blotted out by the canopy.

Some of the birds on the river look very familiar. The neotropical cormorant (Tupi, *bigua*; Portuguese, *mergulhão*), for instance, occurs in small flocks, and like cormorants the world over spends much of its time with wings untidily spread to dry, reminding one of the shags on rocky Scottish coasts. Little martins perch on the midstream rocks and hawk insects over the water, while sandpiper-like waders run about where there are suitable shallow areas of the banks. Others, such as the hoatzin and the snake-bird, look very exotic. The hoatzin is known in Brazil by the rather romantic name of *cigana* ('gypsy woman'). It is about the size of a pheasant, with a light blue face, red eyes, and a tall, untidy crest of pale chestnut feathers, rather like a dishevelled punk hairstyle. Ciganas live in small flocks in the riverside thicket in which they crash about clumsily. Their diet is restricted to leaves, particularly those of the riverside aroid, the *anhinga* (*Montrichardia arborescens*). These are fermented in a large crop, producing the

◀◀ White-winged swallows (*Tachycineta albiventer*) perched on river rocks. (*Dudu Tresca*)

▲ Neotropical cormorants or *biguas* (*Phalacrocorax olivaceus*) are common on the rivers of Maracá. (*Dudu Tresca*)

▼ A hoatzin (*Opisthocomus hoazin*) in dense riverside vegetation.

▲ *Jabiru* storks (*Jabiru mycteria*) on their nest. (*James Ratter*)

► A mixed flight of great and snowy egrets and a little blue heron. (*Casmerodius albus*, *Egretta thula* and *Florida caerulea*). (*Haraldo Palo*)

birds' characteristic unpleasant odour. Uniquely amongst living birds, *cigana* chicks have claws on their wings which they use to scramble about the riverside thicket. With so many unusual features, it is not surprising that the *cigana* is classified in a family of its own. Traditionally it was thought to have affinities with the fowl-like Galliformes, but recent studies show that it is probably more closely related to the cuckoos. The snake-bird has obvious affinities with the cormorant which the Indians recognised by calling it the *biguatinga* ('white cormorant', because of the light-coloured head and neck). It swims with its body completely submerged and only the long neck protruding above the water, thus looking like a snake. Swimming over any distance is slow and laborious, presumably because of the water-resistance of the submerged body. Like the *bigua*, the *biguatinga* spends much of its time standing on rocks or riverside trees drying its outspread wings.

Two species of stork are common and are prominent birds at the water-filled borrow-pits alongside the causeway which leads from the Maracá Channel to the station. Most spectacular are the giant *jabiru* storks which are generally found in pairs. The *jabiru* is a great heavy-looking, long-legged bird, standing about 1.2 metres tall, with a beak about 30 centimetres long. The plumage of the body and wings is a creamy white, and the beak, head, upper part of the neck and legs are black. The neck is very thick, perhaps due to dilation of the crop, and there is a broad scarlet band at its base. The *jabiru* is one of those birds which is so large that take-off is extremely cumbersome and difficult. *Jabirus* build huge platform-like nests of sticks high in the boughs of big spreading riverside trees such as *Ceiba pentandra*, and stand like sentinels on them as one's boat passes along the river. Sometimes their dignity is spoiled by noisy parakeets which find the underside of the *jabirus*' nest a convenient place for their own nests. The wood-storks are somewhat smaller and lighter than the *jabirus* and are gregarious. They are white birds with slightly down-curved blackish-grey bills and have heads, necks and long legs of the same colour. Their heads are naked, like the *jabirus*', and presumably because of this they have the vernacular name *cabeça seca* ('dry head'). *Jabirus* and wood-storks are often seen wheeling gracefully in the thermals high above the river, the latter species sometimes in quite large groups. Both are notable opportunists and are active during the dry season in catching fishes stranded in drying-out seasonal pools.

The ornithologists of the Project recorded no fewer than thirteen species of the family Ardeidae (herons,

▲ Rufescent tiger bittern (*Tigrisoma lineatum*). (*Dudu Tresca*)

◀ A sun-bittern (*Eurypyga helias*) proudly walking up a palm trunk. (*Haraldo Palo*)

egrets and bitterns), reflecting the richness of the aquatic ecosystem. The most abundant of these are the great egret and the much smaller snowy or cattle egret, both with pure white plumage and bright orange bills. Other noteworthy species include the rufescent tiger bittern (*soco boi*) which has very different plumage as a juvenile and an adult. The juvenile has 'tiger-stripes', while the adult is shown above. The call of the *soco boi* is a most distinctive sound, like the roaring of an ox.

The sun-bittern (*pavãozinho*, 'little peacock') is a rather slender, bittern-like bird with beautifully patterned wings. It is sometimes seen flying over the water or feeding at the water's edge or along small forest streams. The photograph above shows its proud, stalking gait. It preys on small fish and other aquatic animals which it spears with its sharp bill. Like the *cigana*, it has so many unusual characteristics that it is placed in a family of its own, the Eurypygidae.

The ornithologists at Maracá found five species of kingfisher, known in Portuguese as *martins pescadores* ('fishing martins', I suppose because, like martins, they live in burrows in river-banks). The largest, *Ceryle torquata*, is an enormous kingfisher, some 42 centimetres long with a dagger-like 8-centimetre beak, and looks rather like a rakish woodpecker. The other four belong to the genus *Chloroceryle* and form a series of pretty, almost identical, green and buff kingfishers differing from one another only in size. The largest (*C. amazona*) is 29 centimetres long and the smallest (*C. aenea*) a mere 12.5 centimetres, of which the beak accounts for nearly 3 centimetres. These four related species are probably specialized to eat fishes of different sizes so that they do not compete with each other – a system to minimise competition and maximise the utilisation of food resources.

A small species of tern and the tern-like skimmers are quite common on the river. In *Brazilian Adventure*, Peter Fleming gives interesting descriptions of the river-life along the Araguaia. He comments on the skimmer:

> The gulls included . . . a graceful, noisy black and white gull with a red bill. One day I saw one of these do a thing I had never seen before; it drank as it flew. It was skimming low, and not very fast, over the still water; its beak was open, and the lower tip just touched the water, scoring a thin unbroken furrow on its surface while the water ran up into the gull's throat.

This is an elegant description of the skimmer's mode of feeding. We now know that it is not drinking but

▲ The head of a submerged Spectacled caiman or *jacaretinga* (*Cayman crocodilus*). (*Haraldo Palo*)

► A baby caiman *jacaretinga* (*Cayman crocodilus*). (*Mark O'Shea*)

running its extended mandible through the surface of the water to catch tiny fishes and crustaceans.

One also sees various species of duck, including the big, heavy, dark Muscovy, ibises of a variety of types and colours, long-legged lapwings with the onomatopoeic name of *quero-quero*, and ospreys just like those of the Scottish Highlands. One could continue at great length about Maracá's many species of aquatic birds, but unfortunately space does not allow it, so let us move on to the reptiles.

Two species of crocodilian (*jacarés*) were recorded at Maracá, the relatively abundant *jacarétinga* ('white *jacaré*') and the less common *jacaréassu* ('great *jacaré*' – *assu* is an augmentative suffix in Tupi). The former is a small species, reaching 2.5 metres at the absolute maximum, but generally considerably less, which was commonly seen basking in the sun on the banks of the borrow-pits at the sides of the causeway to the station. The *jacaréassu* (black caiman), however, is a much larger animal which can reach 6 metres in length. I never saw a large specimen at Maracá, but they were observed by some members of the Project and were reputed to live in the mouths of the tributary streams running into the Uraricoera. The swimming posture of the two species is strikingly different: the *jacarétinga* swims with only the head above water, whereas the *jacaréassu* seems a much more buoyant creature and cruises with the top of the whole length of the back and tail above water. Big *jacaréassus* can be dangerous and are certainly not desirable bathing companions. The *jacarétinga*, on the other hand, is a fairly innocuous animal, although females are aggressive when they are defending recently hatched young. I once met a Brazilian countryman with a badly scarred neck caused by an attack from a female *jacarétinga*. This occurred when, as a boy, he jumped into a pool where she was swimming with her young. The principal food of *jacarés* is fish, and one sometimes sees them with piranhas, surubim or other species held in their mouths.

Anacondas are very aquatic animals, spending much of their time in the water. They were recorded as common by Mark O'Shea, the Project's herpetologist. As is well known, they grow to an enormous size, with at least one reliable measurement of a specimen 37 feet long. This giant was seen in Goiás (Central Brazil) in 1840 by the Scottish botanist George Gardner and had died after swallowing a full-sized horse. There are unauthenticated records of even larger specimens: the famous Colonel Fawcett shot one about 62 feet long on the Rio Acre in 1907, and tells of another 80 feet long killed by the Brazilian Border Commission on the Rio Paraguai! The largest seen on Maracá were a fair-sized specimen, of possibly 4 or 5 metres, photographed by the INPA *técnico* Joselias, and an enormous specimen observed on separate occasions in the Maracá Channel by Amazonas and Guttemberg. Amazonas was the first to encounter this giant snake and wanted to investigate it more closely, but his wife insisted that they leave the scene as fast as the boat could go. He later described it ominously as *um sucuri do tipo que come gente*, that is, an anaconda of the type which eats people! Apart from the last-mentioned item, anacondas normally feed on

South American green iguana (*Iguana iguana*). (*William Milliken*)

birds, small mammals, caiman, frogs, etc. I once saw a 4-metre specimen in the Pantanal do Mato Grosso which had caught a great egret. The other egrets of the flock stood about oblivious of the incident, showing, according to my Brazilian companions, about as much interest as São Paulo pedestrians do when one of their number is killed in the traffic!

A total of five species of turtle was recorded from Maracá Island, and individuals were often to be seen sitting on rocks in the rivers. Two of the species are of the primitive type where the neck folds sideways under the shell, instead of retracting in the more normal tortoise-like manner.

Big green iguanas with comb-like crests are sometimes seen lying along the boughs of riverside trees. When disturbed, they simply jump off to fall with a great splash in the water, there to swim out of harm's way.

When I first visited Amazonian Brazil, well over twenty years ago, one of the animal groups which most impressed and fascinated me by its diversity and abundance was the fishes. During one of my early visits I was very fortunate to have Dr Rosemary Lowe-McConnell, the renowned tropical fish expert, as companion and mentor. The conversations I had with her then stimulated my interest and, together with my own amateur observations and innumerable conversations with fishermen and other country-folk, form much of the basis for the following fishy notes.

The diversity of the fish fauna is enormous: almost exactly one half of the world's 5000 species of freshwater fishes come from the Amazon river system! They range from a host of diminutive species, such as the tetras of our aquaria, to huge 3-metre catfish and even powerful bullsharks which ascend the river system from

▶ Three species of Piranha, black, red and yellow – the easiest fish to catch in Maracá's waters – and, above, a vegetarian *pacu*. (*James Ratter*)

▶▶ Guttemberg Moreno de Oliveiera, manager of Maracá station, showing the fangs of a *pirandirá*. (*James Ratter*)

▶ Teeth of the black piranha. (*James Ratter*)

the Atlantic right into Peru. The most important fishes are the characins and the catfishes, while the carp family (Cyprinidae), so familiar and important in Eurasia, is completely absent from the native fauna.

The characins are a group which have diversified into a host of niches and range in size from tiny tetras to big fruit-eating forms weighing 30 kilograms. They have adopted a great range of feeding habits, from toothless detritus eaters, through vegetarians, to highly predatory forms such as piranhas. Most of the species are highly esteemed as food-fishes, but they are extremely bony. For the squeamish, struggling with the pin-cushion-like body of a piranha or a *pacu* is a never-to-be-forgotten experience. On one occasion I was invited to dinner when the principal dish was *dourado*, a big predatory characin, not unlike a salmon in appearance and much appreciated as both a sporting and a food-fish. I remember contemplating the gastronomic challenge, my eyes alighting on a particularly vicious line of lateral forked bones, each one of which seemed to give me a derisory V-sign. One solution to the problem is often employed in the culinary preparation of the smaller characins such as piranhas and *curimatãs* (the latter a toothless detritus feeder, much appreciated as a food-fish). A close series of deep parallel cuts is made almost vertically right along both sides of the body (which is said to be *lanhado*) and these so fragment the bones that the pieces can be swallowed without ill effects – or so I am assured.

The legendary piranhas are the most famous characins, and also amongst the most common. The name is a contraction of *pira* (Tupi, 'fish') and *çainha* ('tooth') and is pronounced 'pee-ranya'. There are many species of piranha, the most notorious of which are black piranhas (which sometimes reach over 40 centimetres in length) and the somewhat smaller red piranhas (to about 25 centimetres) (see the photograph on page 94). Black piranhas are the more common species in black and clear water, while red piranhas are characteristic of white waters. However, the two species sometimes occur together, as at the Ilha do Bananal in southern Amazonia. At Maracá black piranhas were the common species of the Uraricoera River, while red piranhas occurred in some of the side streams around the swampy areas.

Piranhas have razor-sharp teeth whose cutting-power is astonishing. On one occasion I saw a red piranha bite right through the shank of the strong steel cod-hook which had been used to catch it. It is hard to understand how a creature only 20 centimetres long had such power in its jaws, but my piranha's feat was in no way exceptional: in his authoritative book, *The Fishes and the Forest*, Michael Goulding states, 'The snapping bite of a large black piranha is so strong that a large individual could easily amputate a man's hand at the wrist in two or three bites.' There are many contradictory stories about the danger posed by piranhas, and the differences in their behaviour seem to depend on how hungry they are. Many years ago we swam with the native *caboclos* (backwoodsmen) among shoals of black piranhas in the Rio Suiá-Missu in north Mato Grosso, but these same *caboclos* refused to wade in the shallows of nearby floodplain lakes because of the danger of piranhas. It was the height of the dry season and the lakes were teeming with starving black piranhas which had been cut off from the river by the seasonal fall in water level. The piranhas of the river and lakes belonged to the same population, but availability of food made their behaviour very different. I encountered exactly the same situation during the dry season at Bananal Island in southern Amazonia where we bathed amongst red piranhas in the main course of the Rio Araguaia, but were told that they were extremely dangerous in the floodplain lakes. In fact the wardens of the Reserva do Araguaia had a joke about this: they claimed that to wash when on inland patrols on the island, one had to fill a basin with a lightning dip and then, before starting ablutions, check there was no piranha in it. One recruit smiled ruefully over the joke: he had been given the instructions but had attempted to

Pirapitinga and *pirandirá* – contrasting characins. (*James Ratter*)

A typical catch. Project members were allowed to fish, to add variety to the station's food. (*James Ratter*)

A sting-ray, one of the most feared fish in Maracá's rivers. (*James Ratter*)

A small *pirarara*, a cat-fish that can grow to 50 kilograms. (*James Ratter*)

wash dishes in a stream and received a nasty bite on the hand. People certainly are attacked and sometimes killed: the father of the manager of a hotel in the Pantanal where I stayed in 1986 had been killed by piranhas a fortnight previously, and the secretary of the Ecology Department of INPA told me of an attack close to the station when a small boy bathing in a stream was emasculated. Fortunately there was only one incident with piranhas on the Maracá Project (see page 27).

Piranhas seem to be omnivorous. As well as fish or other flesh, they eat seeds and fruits of the riverside vegetation (and kitchen scraps when they can get them). But some close relatives of the piranhas are specialists in eating vegetable foods and have their dentition modified for masticating hard fruits. These are the *pacu*, *pirapitinga* and *tambaquí* (see the photographs on pages xx and xx). The latter two grow to a great size and their molar-like teeth are capable of crushing the hard nuts of the *Astrocaryum* palm.

The family Cynodontidae ('dog-tooth' family) contains the most alarming-looking predatory characins. They are elongated, slender, hyperactive fishes with fantastic fangs: those on the lower jaw are so long that there are two slots through the upper jaw to accommodate them. The cynodont which occurs at Maracá (see the photograph on page 94) is the *pirandirá* (Tupi, 'bat fish' – presumably because the malevolent, long-fanged face is reminiscent of that of Dracula-like bats). It is about 45 centimetres long and seems to be common; one of the rapids on the Maracá Channel is called the Cachoeira de Pirandirá.

There are many families of catfish in the Amazonian system and the species range from small fishes of a few centimetres to monsters up to 3 metres long and weighing hundreds of kilograms. The largest of these is the *piraíba* ('tree fish'), but there are other enormous species such as the *surubim*, the *jaú* and the *pirarara* ('macaw fish', because of its colouring). Although they have only tiny teeth, some of these big fishes have a reputation for violence. The *pirarara* (see the photograph on the left) is particularly feared, as it has the habit of fastening tenaciously on to the arms or legs of swimmers. In 1980 when I visited the Carajá Indian village of Macaúba in the north of the Ilha do Bananal, one had recently drowned a boy of about ten by seizing his arm and dragging him down: a child is no match for a 50-kilogram fish. The motive for these attacks is difficult to understand as the *pirarara* attacks animals far too large for it to swallow, and does not have teeth capable of tearing pieces off its prey. According to Brazilian country folklore, the fish buries its victims in the mud of the river-bottom and then eats the flesh when it is soft and putrefying. However, it is rather difficult to believe that the *pirarara* is capable of such premeditation, and the tale can probably be regarded as a fisherman's tall story – appropriately, I heard it, more than once, in bars on the Rio Araguaia. The huge *piraíba* also has a bad reputation, and Alfred Russel Wallace in his *Travels on the Amazon and Rio Negro* mentions that in the Indian villages on the Rio Negro the children bathed within enclosures of stakes to prevent its attacks. Some of these catfishes, such as the *surubim* and the *filhote*, are excellent eating, and, unlike the characins, their flesh is practically boneless. However, they are avoided by many Amazonian country people because of a superstition that eating their flesh causes leprosy. The commonest large catfish at Maracá seems to be the *surubim*, but on my last night on the island Guttemberg caught a fine 40-kilogram *jaú*. On opening it we found that the gut was completely empty and that it was a female with a huge ripe roe like that of a big cod. Since then he has caught a *piraíba* with an estimated weight of 200 kilograms. The huge fish was taken by the barrel-fishing technique in which the fish is exhausted by having to tow a large float (the 'barrel'). Even with this, however, Guttemberg did not succeed in landing it, and it was discovered and secured by a neighbour the next day, by which time it was completely tired out.

All the catfishes so far mentioned belong to the family Pimelodidae and have very tiny scales which are not apparent, so that the Brazilian name for them is *peixes de couro* ('leather-skinned fishes'). Some other catfish families, however, are covered with heavy, bony armour. One of these families is the Doradidae of which large species were caught at Maracá (see the photograph on this page). They have impressive bony spines in front of the pectoral and dorsal fins. As I write, I am looking at one of these spines taken from the front of the pectoral fin of a 70-centimetre doradid. It is 12 centimetres long and has sharp teeth along both edges, so that it looks like a miniature version of the beak of a sawfish. It represents a modification of the first fin-ray, and its base, which must articulate on to the pectoral bones, looks like a mammalian vertebra. The gut of this doradid was filled with large, hard, palm fruits.

The *acarí* (family Loricariidae) is another armoured catfish and spawns in burrows in the river-bank during the rainy season. Large numbers of these burrows are made close together so that when they are exposed by the fall of water in the dry season it looks as though the bank is inhabited by a colony of sand-martins. The trunk of one of the common trees of the Amazonian region, *Minquartia guianensis*, is pocked by a pattern of holes, and, alluding to this, its vernacular name is *acarí-coara* ('*acarí*'s burrow'). To me this is a jewel of knowledge since I had known the vernacular name for years without understanding its meaning, but to the reader it is no doubt an utterly inconsequential remark! The *acarí* is a relatively small fish, reaching at the most 20 centimetres, and is sometimes seen in tropical aquaria. It has a downwardly directed mouth, well adapted for sucking algae from rocks, and, like the doradids, looks rather like an ancient armoured relic of the Palaeozoic era. Like many neotropical fishes it can breathe air, and the surface of stagnant waters is often in continual disturbance by a great number of *acarí* rising vertically to gulp air.

Efrem Ferreira from INPA studying a Doradid catfish in one of the station's laboratories. (*William Milliken*)

There are a number of species of sting-rays in the Amazonian river system, and in the other great Brazilian rivers. Sting-rays, like sharks, are cartilaginous fishes and are a marine group which has adapted to freshwater life. For such exotic-looking animals they are surprisingly abundant: sometimes, when the water level of the Santa Rosa Channel was low, they were swimming in such numbers as to make the bottom look like an undulating carpet. They even penetrate into quite small streams, and the first I ever saw was a large specimen, about 80 centimetres across, caught in a shallow stream no more than 6 metres in breadth. They have the habit of burying themselves in the sand or mud of the shallows with just the eyes protruding and thus are often accidentally stepped upon. When this happens, they slash with their tails and inflict severe wounds with long, barbed dorsal spines. The spines carry a venom which causes agonising pain, so that a wound from a ray is remembered for life. Apparently, however, the venom is destroyed by heat, so the pain is much alleviated by extremely hot compresses. This treatment is not widely known, but there are highly indelicate local remedies which possibly owe any effect they may have to raising the temperature at the site of the wound – for instance, getting somebody to urinate on it. The venom on detached spines is said to last for many years, and according to hearsay they were used in police interrogation tortures during the time of the military regime.

Perhaps the foregoing conveys a little of the paradise which Amazonia represents for the ichthyologist. Reading over these pages, I am struck by the apparent malevolence of neotropical fishes with their armaments of sharp teeth, poison-spines and other fiendish contrivances – so different from the toothless innocence of our own dear roach and carp. But no doubt there is too much emphasis on violence, as in the popular press, and not enough on the beauty and interesting behaviour of many of these fishes. For those who want to learn more of Amazonian fishes from a real authority, I recommend Michael Goulding's wonderful books, *Amazon – the Flooded Forest* and the much more technical *The Fishes and the Forest*.

Sadly, the most obvious invertebrate of the river occupies in terms of human irritation more or less the niche of the *carrapato* on land. The *pium* is a diminutive black-fly (Simuliidae), about the size of a fruit-fly, which breeds in countless millions in places where there are rocks and fast-flowing water: an abundant habitat on the Uraricoera River. The larvae live attached to the substrate, feeding on tiny micro-organisms in the water, but, as in mosquitoes, the adult female has to take a blood-meal before it can reproduce. The bites produce small swellings with little blood blisters at their centres and in profusion cause such irritation that it feels as if the skin is burning. In addition to this irritant nuisance, *piums* can carry river-blindness (Onchocerciasis), a disease caused by a parasitic nematode worm. Both the disease and its simulid vectors were studied by Dr Victor Py-Daniel's team. River-blindness was not found in Maracá Island itself but affected the Indian tribes of the northern tributaries of the Uraricoera River. It does not appear to be an endemic neotropical disease and, as previously mentioned, was probably originally brought to South America by negro slaves from West Africa. Unfortunately it now seems to be spreading in Brazil, possibly because of infected military personnel who have served on the Roraima frontier returning to their homes in other areas and spreading the disease via the local *piums*.

Drs Cecília Volkmer-Ribeiro and Maria Cristina Mansur studied the sponge and mollusc faunas of the aquatic habitats (see page 109). Five species of encrusting sponge were found in the seasonal lakes and streams of the island, but only one in the Uraricoera River. The last probably belongs to the genus *Oncosclera*, characteristic of fast-flowing and turbulent waters, and was rare and only collected once. On the other hand, the sponges of the seasonal ponds are common and encrust any submerged object suitable as a substrate, such as fallen *buriti* palm fronds. The silica content of the water is much reduced since it is utilised in the construction of the skeletons of the abundant sponges. Entering the water of these seasonal ponds causes an uncomfortable itch brought about by irritation of the skin by loose sponge spicules and this persists for a number of days. Dr Arno Lise, who collected sponge specimens from the centre of the ponds, was the leading sufferer from this uncomfortable complaint. The Uraricoera River flows over the mineral-poor Guyanan shield and its calcium content is rather too low for molluscs to develop their shells. The Malacofauna is therefore poor in number of species and individuals: only three species of bivalve mollusc and one gastropod were recorded from the river, while molluscs were absent from the seasonal ponds, and only one gastropod species occurred in the flooded borrow-pits. This is in striking contrast to areas like the Pantanal where the superabundant apple-snails provide an important part of food-chains.

I HOPE that these pages have given the reader some idea of the beauty and wonder of the Uraricoera River. The great American explorer Hamilton-Rice went so far as to claim that 'in no part of South America does another such river exist'. It would be nice to report that the river is still in as good condition as at the end of the Project three years ago, but unfortunately because of recent developments this is not the case. For many years there has been some gold extraction in the region, but towards the end of 1987 a major gold-rush began, principally upriver of the island and in the Yanomami Indian lands to the west. Great numbers of miners (*garimpeiros*) poured in by air and started work in an atmosphere of gold-fever (see page 116). The gold is extracted from river sediments which are worked using moored *balsas* (rafts). Each raft has a crew of about four to five *garimpeiros* and carries a pump which is used to suck the deposits from the bottom of the river; the hose on the bottom is directed by a diver and frequently work continues day and night. In 1988 there were numbers of these *balsas* on the Maracá Channel quite close to the station, but fortunately gold production was insufficient to cover their expenses and they moved away. The nearest *balsas* to the station now are on the Santa Rosa Channel, with one at the Cachoeira de Fumaça, almost at the northernmost point of the island, and another further upstream at the Cachoeira da Onça. Still further upstream, above the island, there are reportedly no fewer than 800. All have been transported by air to specially made airstrips which are extremely busy. The largest aircraft used are DC3s, and to fit into their fuselages the steel floats of the *balsas* have to be cut into three pieces which are later welded together again. As can be imagined, fortunes have been made, not least in transport: one group bought a new Bell helicopter and made profits which paid for it in one month. Many lives have been lost, not only in plane crashes but in accidents on the turbulent rivers. One group of *garimpeiros* trying to ascend the Maracá Channel encountered the wreck of a previous crew at the Cachoeira de Pirandirá: a boat was half-submerged against a rock in mid-stream and a body was being tended by king vultures. The *garimpeiros* who related the story wisely turned and came back!

The gold-rush has had profound effects on upriver Indian communities. Places which were in the middle of nowhere three years ago now have airstrips, shops, bars, and the type of bustle and high jinks associated with similar communities in Western films. Inevitably pollution of the river has occurred; for instance, at the Maiongong Indian village at Uaicas a slick of diesel oil was seen which was so long that it took a full hour to pass down-river. The Maiongong are accustomed to a very clean environment and no longer use the river for drinking water or even for washing. Probably the most dangerous, and least seen, pollutant is mercury, which is used in the extraction of gold and escapes into the aquatic environment. Mercury passes along the aquatic food-chains and accumulates in the end-predators such as fish-eating birds. The effects can be serious; for example, studies in other parts of Brazil have shown mercury approaching lethal levels in kingfishers.

The disturbance is mainly upstream of the island, and no doubt the Maracá and Santa Rose Channels still appear unchanged, but mercury will be invisibly and insidiously creeping along their food-chains. It is difficult to predict the long-term effect of this contamination; hopefully mercury pollution will decline as the gold-rush ends, and the ecosystem will prove itself resilient enough to withstand its effects.

It now seems almost obligatory to end writings such as these with tidings of impending disaster. This leaves the reader in a state of pessimistic gloom, so different from the exhilarating picture of magic new worlds created by the works of the great nineteenth-century naturalists such as Bates and Wallace. But anyone sufficiently interested to read to this point will be all too familiar with the problems of the destruction of the tropical rainforests, the greenhouse effect, the hole in the ozone layer, and dozens of other ecological issues, so there is no point in rehearsing them here. Clearly the importance of these matters is paramount, and responsible stewardship of the biosphere is Man's greatest challenge for the future. In Amazonia there are still huge unspoiled areas, and with the right scientific, economic and political will, a combination of conservation and rational land-use could maintain a biological paradise in perpetuity. Let us hope that Man achieves this, so that in the future our descendants can appreciate the marvels that we enjoyed.

CHAPTER FIVE

Results

THE main invitation from SEMA, the Brazilian Environment Secretariat, was to produce a complete ecological inventory of the enormous Maracá Reserve. This was clearly an impossible task, but the Project can certainly claim to have made the most intensive study ever of a rainforest area at the northern edge of the Amazon basin. The value of this survey was enhanced by the fact that Maracá lies on two important biogeographical boundaries: between Amazonia and the Guianas, and between natural forest and open savanna grasslands. It was also important that the rainforest at Maracá is essentially marginal in character, with a fairly low rainfall and a number of deciduous tree species. Marginal tropical forests cover vast areas and are therefore critical ecosystems, often threatened with destruction. But they have been less studied than wetter, taller rainforests.

The Ecological Survey was a kaleidoscope of individual research efforts. Each scientist made minute and often highly specific observations; but every contribution added another piece to the ecological jigsaw. Jim Ratter, as leader of this programme, welded all this research into the desired ecological survey of Maracá. His own contribution was in the botanical survey – the largest of all the Project's findings – which is described in the chapter on the land environment.

As, in sheer weight, insects account for about 50 per cent of a rainforest's animal biomass, it was therefore appropriate that no fewer than fifty-two scientists made entomological studies, under the direction of Drs José Rafael and Victor Py-Daniel, respectively of INPA's entomology and medical entomology departments.

By the time of the second review conference in London, many of the groups of insect specimens had been sent to world experts for determination. This confirmed that over 200 new species had been discovered, all new to science and described and named for the first time. The final count for new species resulting from the Maracá Project will certainly be over 250.

It is worth recalling that this was *not* a collecting expedition. The Project's 30,000 insect specimens were gathered by some trapping close to the station and on river trips by the medical entomologists; but this was done as part of wider research and was not intensive collecting. The fact that so many discoveries could be made, almost casually, is a measure of the amazing biological diversity of rainforests. Had the Maracá Rainforest Project collected vigorously all over the 110,000-hectare island, there would certainly have been many more discoveries of new species.

One of the largest insect groups was the Hymenoptera, which means those with membranous wings. This order of insects includes bees, wasps, sawflies and ants. The entomologists collected 145 species of Hymenoptera, in sixteen genera. Of the 100 species of Psocoptera (winged lice, etc.) collected, no fewer than sixty-six were new discoveries. There were twelve new species of Homoptera (uniform-winged bugs such as cicadas or green-fly), three of Trichoptera (hairy-winged creatures, including caddis-flies), and ten new

Fronds of *buriti* palm (*Mauritia flexuosa*). (*Dudu Tresca*)

species of Pipunculidae (a group of small black flies) out of eighteen collected.

The butterfly collectors Olaf Mielke and Mirna Casagrande from the south Brazilian state of Paraná netted over 2700 specimens of 458 species. They worked manually, with traditional butterfly nets and also with cylindrical traps 50 centimetres in diameter, baited with banana skin fermented in sugar-cane juice. They came to the island twice and found that collecting was poor in May, at the end of the dry season, but much better during the rains in August. Olaf and Mirna's dazzling collection of butterflies and moths, all beautifully mounted, is now in two Brazilian institutions, just as all the other insects are lodged in collections in their country of origin. They managed to collect both male and female of a completely new species – a rare event in the world of butterflies. It has been named *Pythonides maraca*.

The Project's spider expert, Professor Arno Lise from the Zoobotanical Institute in Rio Grande do Sul, was also deeply disappointed when he first came to Maracá during the dry season. He found almost no spiders and cannot explain where they were hiding. This changed dramatically when he returned later in the year. Arno Lise then collected spiders of some 275 species and thirty-two families. His forest collections yielded a remarkable twenty-two new species and one whole new genus or even family of spider. More new discoveries are anticipated from his savanna material. A meticulous scientist, Professor Lise is also a fine photographer. His pictures of spiders' webs show their intricate geometric beauty and their diversity. The webs range from tiny strands to cocoon-like tunnels, from concentric rings to the gigantic edifices of colonies of social spiders. These tiny creatures build metropolises that stretch for several metres up the sides of trees.

Pedro Linardi from Minas Gerais studied ectoparasites of rodents and marsupials. He first had to catch the host animals, which he did using traps baited with fermented banana skins, manioc, seeds of corn, sunflowers and peanuts, and a nasty mixture of rodent viscera and Scotts Emulsion. He collected 1800 ticks, lice and fleas from the skins of fifty-one wild rodents and three marsupials. All but three of his parasite species were recorded for the first time in Brazilian Amazonia, and two species of ectoparasite were new to science: an acarid (tick) and a biting louse.

Gilberto Righi from the University of São Paulo is one of Brazil's leading experts on earthworms. He found Maracá to be an ideal location for his work, rich in material and diversity of habitat. There are large quantities of earthworms near the watercourses and in the island's swamps and stands of *buriti* palms, where they are dug up by peccaries and foraged by birds. Although present in such large numbers, Maracá's earthworms come from only a dozen species. One of these was new to science, a 37-millimetre, colourless worm found in damp black soil beside a swamp. Righi has named his discovery *Glossodrilus motu*.

Maracá is swarming with ants. At the edge of the station clearing, where the main trail enters the cool gloom of the forest, there is a big underground nest of leaf-cutter ants, *Atta cephalotes*. At ground level this ant-hill looks like a pink sand-pit in a children's playground, but its labyrinth of tunnels stretches for many metres below the surface. We have already described the familiar and fascinating sight of long columns of leaf-cutter ants scurrying along smooth trails, each carrying a slice of leaf held aloft like a green sail.

One of the Project's ant specialists was British entomologist Forbes Benton, who has worked for years at the Brazilian government's cocoa research institute in Bahia. Forbes spent several months on Maracá, observing ants and particularly the tiny fly parasites that prey on them. He noticed swarms of small flies of the family Phoridae that jab at the leaf-cutter ants as they re-enter their ant-hill. This jabbing is done by females apparently trying to rub eggs off on to the backs of *Atta cephalotes*. The swarms of parasitic flies (some of which were species new to science) attack at dawn and dusk. The ants seem to have reacted by changing their foraging pattern. At the start of the Project, columns of leaf-cutters were visible every day. But by the end of the dry season, they had switched completely to nocturnal activity: their trails were deserted during daylight and the thousands of ants with their leaf fragments could be seen only at night by torchlight.

Forbes Benton also studied army ants, *Eciton burchelli* and *Eciton hamatum*, that are present all over the forests of Maracá. The seething mass of these deadly insects has an awesome fascination. Their raiding columns scurry with ruthless determination over or under every obstacle on the forest floor. These columns are only about

20 centimetres wide, and you can step over one of them without being molested. At the head of each column – which can be hundreds of metres from their bivouac – the army ants fan out in attacking formation. Woe betide any creature, even an immobilised human being, that does not run away from the attacking phalanx of fearless and aggressive *Eciton*! They kill everything by sheer weight of numbers and reckless bravery.

Forbes Benton once watched army ants climb a tall dead tree to attack a colony of the common *Camponotus* ants. During the skirmish, many *Camponotus* fell or jumped to the ground. As they ran off, they were buzzed by *Diocophora* parasitic flies, which hovered above the army ants. The flies followed every movement of the zigzagging *Camponotus* until they succeeded in jabbing their eggs on to the hapless victims. On another occasion, I was with Forbes when we watched *Eciton burchelli* swarm up a tree that contained a large carton-like nest of black *Azteca* ants. There was a brief, fierce battle, with dead ants falling off the tree trunk. The *Azteca* won, thus preventing the army ants from climbing further to plunder a tempting nest of *Polybia* wasps higher up the tree. When the army ants decided not to pursue this skirmish, they left simultaneously, sweeping off down the tree with military precision as swiftly as they had arrived. They moved on and swarmed up the next tree to devour all its grubs and insects.

During the battle between the two ant species, Phorid parasites jabbed at the *Aztecas* while they fought off the army ants. Back at the *Eciton* bivouac, another species of Phorid parasite was lying in wait. Forbes Benton noticed that only the males of this species had wings: the males therefore had to airlift successive females (who outnumbered the males) from each ant bivouac to the next.

Forbes Benton is a dedicated entomologist. He therefore made tough journeys up the rapids and waterfalls of the Maracá Channel, to the far western end of the huge island, collecting Phoridae with malaise nets all along this difficult route. He collected at twenty-seven sites, for a total of 125 days, and nine of the twenty-five species of male *Dohrniphora* flies in his catches were previously unknown.

Many ants establish symbiotic relationships with plants. The small *Azteca* ants that had fought the army ants also protect *Cecropia* trees against the ravages of leaf-cutting *Atta* ants. Some species of *Cecropia* attract *aztecas* by providing attractive homes in the stems of their branches. Other types of *Cecropia* are myrmecophobic: they repel all ants, apparently by secreting a poison. Pierre Jolivet, an enthusiastic French entomologist, studied ant–plant relationships and found that *Azteca* ants were generally quite good at defending their host trees against defoliators. But Jolivet recorded one instance when this protection failed.

> One tree of *Cecropia palmata*, situated very near a huge *Atta* nest, had suddenly been attacked by hordes of *Atta* workers and the invasion lasted just three days, the time needed to devour the tree's entire foliage. The *Azteca* ants withdrew within the stems in cowardly manner and stayed hidden. After the raid the *Azteca* emerged, life returned to normal and the leaves of the *Cecropia* rapidly grew again. The *Atta* were soon punished, since a burrowing animal, probably an armadillo, destroyed their nest in one night.

These are just a few examples of the countless relationships between the millions of species that inhabit a tropical forest. Some relationships are benign and mutually beneficial, others are hostile – the law of the jungle at work in the constant struggles for food and reproduction.

Teams cutting trails through the forest or clearing sites for camps along the river's edge were always on the look-out for wasps and particularly for hornets. Members of the Project were occasionally stung, but luckily never by the dreaded hornets whose sting is dangerous and extremely painful. Our revenge on the wasps was exacted by Tony Raw, a highly experienced entomologist working at the University of Brasilia. Tony collected wasps and their nests in many parts of the rainforest, but also under bridges and in plastic pipes in the station itself. He recorded thirty-five species of social (as opposed to solitary) wasps on Maracá Island, and twenty-one of these species had never previously been seen in Roraima. One species was completely new: a 14-millimetre wasp with a golden-yellow body with black and brown markings on its back. This species is now called *Mischocyttarus maracaensis*, and specimens of it are lodged in Manaus, Brasilia and Washington, D.C.

Dr Raw studied the habitats and hunting grounds of

his social wasps. He found that each species tended to make its nest and hunt for insects in a particular habitat – forest clearings, patches of open savanna, near the forest floor or in the canopy. *Polybia liliacea*, for instance, lives in big nests up to 2 metres long near the tops of trees. The thousands of wasps in such a colony rarely descend to ground level, since it would be difficult for a wasp to fly back up to its nest between the forest trees carrying its insect prey. Tony Raw found that wasps were attracted by the artificial clearings created by the Forest Regeneration team, presumably because the larger clearings were full of their prey species. Temperature and humidity were important considerations for different species of wasp in deciding where to build their nests.

Over 100 scientists were engaged in collecting data for the ecological survey, but they were not solely collectors and taxonomists. They also devised a wide variety of experiments. For instance, Forbes Benton wanted to know more about the roles of bees and humming-birds in the pollination of a conspicuous passion-flower vine, *Passiflora longiracemosa*.

He measured when and for how long each reddish-pink flower of the *Passiflora* remained open, using loops of wool and daily observation of individual flowers. Each was found to open for only one day, but with the opening of the flowers in each bunch staggered over a period of days. Further experiments with plastic bags, cotton cloth and chicken-wire revealed how and when each flower's pollen was removed, and the extent to which different creatures contributed to pollination. Forbes found that the small *Tetragona* bee, which scraped pollen into baskets on its hind-legs, was a pollen robber that did not help to fertilise the vine by moving on to the stigmas of another plant. This task was performed by hermit humming-birds, which hovered and fed at a flower cluster before darting on to the next. Another larger bee, of the Meliponinae, stole the plant's nectar by cutting a hole at the base of its calyx tube to get at the floral nectary.

Forbes Benton also observed instances of symbiosis, the fascinating process whereby different species in a rainforest help one another. The nectar of the passion-flower vine attracts various species of ant, and they in turn protect the plant against predators. Forbes watched ants who were enjoying the nectar of one vine fight off foraging leaf-cutter ants. They also snatched individual army ants who came too close to their host. By contrast, another *Passiflora longiracemosa* that had failed to attract such ant guards was demolished by leaf-cutters.

The importance of termites in rainforest ecology cannot be exaggerated. They perform an essential task in breaking down dead wood and vegetable matter, quickly and highly efficiently. Thanks partly to termites, the nutrients in forest litter are recaptured by growing plants. Termites are present in immense quantities: in sheer tonnage, they far outweigh all the larger forest animals put together. They are an important food-source to many creatures, the base of various food-chains. The Yanomami and other Indian tribes eat termites, which are quite tasty and full of nutrition, and could become a food resource for other human beings or their domestic animals. On a negative note, termites have been expanding their numbers hugely, to keep pace with the rate of deforestation around the world. In consuming wood, termites produce methane. They may therefore be one of the causes of the alarming increase of methane in the atmosphere, which is both a potent greenhouse gas and one of the agents damaging the ozone layer.

The termite team on the Maracá Project was led by Adelmar Gomes Bandeira of INPA and Eliana Cancello of the University of São Paulo. They observed the feeding habits of different types of termite, in both forest and in various sizes of clearing. They found fifteen species of the most common genus of termites, *Nasutitermes*, all of which eat wood and occasionally leaves. Three species of *Ruptitermes* (one of which was new) prefer leaf litter or rotten wood. A discovery was that *Syntermes* species specialise in eating only leaves, which they cut up on the ground and carry to their subterranean nests to be stored and eaten later. The *Ruptitermes* do the same. There were roughly thirty-two to fifty-two colonies per hectare of these genera of termites in virgin forest, and sixteen to twenty-two colonies in artificial clearings.

Reptiles and amphibians play an important role in the food-chains of a rainforest. We have seen how the young herpetologist Mark O'Shea was tireless in trapping and catching these creatures, using a formidable armoury of traps and weapons. Mark O'Shea's observations and collections added to records made by herpetologists from INPA. His work raised the number of recorded amphibian and reptile species on Maracá to over 100. Like most of the Project's collectors, Mark rarely killed his specimens: some were released immediately after capture, others after a period of captivity in the station's laboratory where they were carefully studied and photographed. One wall of a laboratory was lined with cages containing a shifting population of frogs, toads, caimans, lizards, hylids, iguanas and, of course, some fearful-looking snakes.

Amid the leaf litter on the forest floor are many species of tiny geckos, including *Coleodactylus septentrionalis* which is only 4 centimetres long, and diminutive frogs. These small animals form the diet of reptiles such as the common striped whipsnake, which hunts by day and moves rapidly, and the aggressive rear-fanged green lora snakes. Mark was pleased with a number of blind snakes, which live below ground, that he caught in his drift fences: some of them were of species not previously recorded in this part of Brazil.

Everyone arrives on Maracá Island at a landing-place with a steep and often slippery mud bank. Between there and the station buildings is the 2-kilometre causeway across wetlands. Where the earth had been excavated to make this road, there is a line of lagoons that were full of caimans during the dry season. The presence of large numbers of these caiman was another sign of how well Maracá has been protected, for they are hunted elsewhere for their skins. During the rains, adult caiman disperse across the plains, but they return to the lagoons as the water falls. Mark O'Shea wrote:

> At the height of the dry season more than a dozen adult caiman could be seen lying, gasping to keep cool, on the banks or in the shallow water while others sheltered under the limited vegetation cover. The author was fortunate enough to be present shortly after a hatch of juvenile caiman early one morning. A torch shone across the water reflected almost seventy pairs of baby caiman eyes and prompted a chorus of tiny caiman voices calling for their mothers.

The station's four buildings were home to many geckos and other lizards. Mark O'Shea could thus add to his collections while relaxing on the veranda in the cool of the evenings. In his own words:

> The presence of lizards provided easy meals for the island's commonest snake, the cat-eyed *Leptodeira annulata*. These snakes could be found on most evenings hunting geckos on the verandah and the author captured eighteen without much effort, one actually crawling under his chair while he sat drinking a *caipirinha*. A good-sized bat population also tempted the occasional tree boa, *Corallus enydris*, into venturing on to the verandah. One never knew what would turn up next, a *Chironius* tree-snake in the lemon tree behind the kitchen, and juveniles of both the rattlesnake and the Fer-de-Lance on the ground directly outside the dormitories. . . . The largest snake encountered within the station compound was a 2-metre common boa, *Boa constrictor constrictor*, captured in the centre of the camp after a heavy rain storm. This specimen had obviously struck at a porcupine, because he had five quills embedded in the roof of his mouth, with one actually penetrating through the top of his snout. These spines were removed and the boa released.

A rainforest contains a wide range of small mammals, even though they are not present in large numbers. There is little food for them on the forest floor apart from fruits and nuts from the trees above. Rodents are important agents for dispersing the seeds of certain tree species. The agouti, for example, is a large rodent with exceptionally powerful jaws and is the only animal able to open the fruit of the Brazil nut tree. Agoutis are thus essential for the survival of that commercially important species.

Our team of small-mammal collectors spread traps on the forest floor and in low vegetation – wherever they could locate burrows, mammal runs (where animals beat a path through the dead leaves and litter), or places where small mammals had been eating or bedding down. At times they had up to 250 traps in operation. Most small mammals operate by night, so the main inspection of traps was carried out at dawn. Adrian Barnett and Aléxia da Cunha set a total of over 4600 'trap-nights' during both the dry and wet seasons, but caught only seventy-two animals. They attributed this low success rate to the sparsity of food on the impoverished soils of the forest floor, which caused mammals to range farther than in richer forests, and also to the fact that most trapping was in the dry season when there were fewer young animals about. Whenever an animal was caught, the team took about two and a half hours to record all data about the site, and then measured and recorded all that they could about the small captive before setting it free. Their work went far beyond taxonomic identification: it broke new ground in discoveries about behaviour. With fewer than twenty full-time small-mammal scientists at work in all Brazil, there is vast scope for further research.

The small-mammal team caught animals representing eleven species from two orders and three families. They also observed or gathered data about a further five species not caught in their traps. These animals ranged from the tiny, shy mouse opossum (*Marmosa murina*), a marsupial that does not have a pouch – its young must cling to their mother as best they can. The many different species of rats and mice recorded included the large marsh rats (*Holochilus brasiliensis*), the fish-eating river rat (*Nectomys squamipes*) and various arboreal creatures such as the singing tree rat (*Dactylomys dactylineus*) which lives high in the canopy and emits a distinctive call. There were also voles in the grasslands on Maracá and porcupines and squirrels living in the trees.

Another important study of animals on Maracá was of primates. Andrea Nunes and her colleagues from the Goeldi Museum in Belém published a handsome booklet on primates of Roraima, with coloured paintings of all eleven species known to occur there. Only five of these types of monkey were seen on Maracá Island. Andrea observed eighty-one groups of primates during almost 300 kilometres of walking through the forest.

The island's most common primate was a large spider-monkey with the splendid zoological name *Ateles belzebuth belzebuth*. These spider-monkeys normally eat fruit, but examination of those on the island revealed that they had been eating a fair quantity of leaves. This might explain their high density of 1.2 groups observed per 10 kilometres. The largest individual *Ateles belzebuth* was a female weighing 9 kilograms, and they moved about in groups of up to nine animals.

Another common primate on Maracá was the red howler-monkey (*Alouatta seniculus*). Individual howler-monkeys weigh almost as much as the *Ateles* spider-monkeys. Everyone who visits an Amazonian forest knows the loud and lugubrious howling of these monkeys. You wake in the night to the noise of what sounds like the cheering of a distant football crowd, rising and falling as it echoes through the forest.

Cebus apella is a lively little Black-capped Capuchin monkey, often seen swinging in large bands high up in the tree canopy. These monkeys sometimes have related *Cebus nigrivattatus* mingled in their groups. They eat everything from fruit and nuts to ants. Lastly, there was a small grey squirrel-monkey (*Saimiri sciureus sciureus*), seen near swampy areas and streams, in troops of between twenty and thirty individuals.

Maracá is a wonderful haven for birds. In places, the rivers and wetlands were like aviaries with every type of bird to delight even the most uninitiated. There was a full range of bird sizes, plumages and behaviour patterns. Ornithologists had previously spent brief periods observing Maracá's rich bird-life, so that the first task of our ornithological team was to record new species. Their observations brought the list of species sighted on this island to over 440. This great diversity was due to the variety of habitats – dry *terra firme* forest, riverine environments, wetlands, and savanna – and also to the fact that we were in a region of overlap between the geologically ancient Guiana hills and the main Amazon basin. One ornithologist, the lively José Maria Cardoso da Silva from Belém, also studied bird behaviour. He entertained other scientists in the evenings with stories of his birds' courtship rituals, in which males hopped along a branch to take their turn trying to impress females with acrobatic displays.

The Maracá Project was fortunate to have two passionately dedicated bat experts, Rogério Gribel from INPA and Fif Robinson from England. Their combined efforts identified forty-eight species of bat, which makes the Maracá Reserve one of the richest areas of the New World in diversity of species of these animals. This is half the total of bat species known throughout the Amazon basin, and compares with a mere fifteen species in the British Isles.

Maracá's bats range in size from tiny fruit- and insect-eating species weighing only 2 or 3 grams, to relative monsters of 100 grams. We have already heard about the big 62-gram fish-eating *Noctilio leporinus*, that skims across the rivers at night and gaffs out fish with claws on its feet. Its sensitive echo-locating system can apparently detect ripples on the water's surface that betray the presence of fish. Fif Robinson discovered that the short fur on this bat's wing and tail membranes is covered in a sweet-smelling oil that helps to waterproof them. This oil is so pungent and its smell so distinctive that anything it touches retains the smell for many days.

Another big bat species is *Pteronotus parnelli*, an insect-eater that weighs up to 32 grams at Maracá – far heavier than records for this species elsewhere. A smaller related species, *P. personatus*, usually forages by night in pairs of males and females. But on one occasion, Fif netted over thirty females in a single location; he cannot explain why these females had congregated.

A widespread family of bats in tropical South America are the leaf-nosed Phyllostomidae: their snouts are flat and taper up to leaf-shaped points. These common bats play an important role in the complicated rainforest ecosystem, since they help disperse seeds or pollen of many plants and trees. Many of those captured on Maracá had sticky seeds caught in their fur or had eaten fruits and were about to drop seeds through their faeces. Various blood-sucking vampires were caught, including *Desmodus rotundus*, the common vampire bat found throughout South America. It has pointed ears and sharp teeth, as would be expected from its gruesome reputation.

The world's most common family of bats are insect-eating Vespertilionidae. Of these, *Myotis* can claim to be the most widespread mammal genus – after man – since species flourish in both tropical and temperate regions in almost every continent. Most of these bats captured on Maracá were small, weighing on average only 5 grams. Another bat family that is familiar on Maracá are Molossidae (free-tailed bats). These creatures cleverly evade their enemies by roosting in cracks within rocks in the midst of the fast-flowing rapids of the rivers that embrace Maracá Island. Whenever we travelled up the rivers and landed on these rocky islands, we became aware of the smell and squeaking of *Tadarida laticaudata*. Their silhouettes could be seen, moving about in crevices of the black rocks. Fif Robinson caught all his specimens of this species in nets slung over swiftly flowing parts of the rivers. One evening, he watched a huge flock of thousands of these bats flying at dusk across the water.

Fif was a tough expeditioner, agile at leaping between slippery rocks in the rapids. He penetrated the remotest parts of the island to collect specimens. His biggest catches were at riverside sites, but he also fixed his traps at different heights in clearings and paths in the forest. He and Rogério generally used mist nets, roughly twice the height of nets on tennis courts and often many metres long. Mist nets are the standard tools of both bat experts and ornithologists, but they have to be constantly monitored, since bats trapped in them struggle violently and can get entangled, distressed and wounded, or can escape. Our bat collectors were therefore active at night, using head torches while they worked to dislodge their specimens. They wore a tough glove on one hand to avoid painful bites, while using their other hand to free the animal. Fif Robinson also deployed a harp trap, a recent invention consisting of very fine and strong tungsten wires strung vertically on a frame 2 metres square. The wires of a harp trap are so thin that they can catch some bats whose echo-location systems are too sensitive to be deceived by mist nets. Surprisingly, bats that fly into harp traps glide down into a canvas collecting-trough underneath, where they quietly go to sleep. The drawback with harp traps is that they are cumbersome and fragile: the expensive tungsten wires can easily snap or sag, so Fif spent many hours each day rewiring his trap.

Bats are vital to the maintenance of the forest ecosystem in many ways. They hold down the insect populations, and they disperse seeds and pollen of many tree species of the Amazon rainforest. Trees which are pollinated by bats have characteristic blossoms: large white flowers that produce plenty of nectar at night, give off a strong musky smell, and grow well clear of branches so that the bats can easily and safely reach them. Trees that are bat-pollinated include bananas, *Caryocar*, balsas and the magnificent giant kapok, *Ceiba pentandra*, whose majestic crown often towered above the canopies of other trees at Maracá.

The bat team wanted to discover which species lived on the island, where they preferred to live and forage, and what they ate. It is difficult to observe bats as easily as ornithologists can watch birds. They must therefore be caught. To avoid killing their specimens, our collectors photographed them (by flash, in the total darkness of the night-time forest) and then carefully measured and weighed them. Particular attention was paid to their forearm and wing measurements. A bat's teeth are important in determining its species, but it is obviously

difficult to observe the teeth of a tiny squirming creature. Fif therefore invented a system he called chewits. He obtained small wood spatulas from an ice-lolly maker and silicon putty from his dentist. A tiny amount of silicon putty was gently pushed into the bat's mouth and its jaws were pressed down or up on to it to obtain perfect casts of the animal's teeth. The bat was then set free, frightened but physically unharmed, having made its contribution to science. This 'chewit' technique has been reported to the Mammal Society and should revolutionise future bat studies. If so, it will save the lives of many of these valuable creatures.

Two lady scientists from the south Brazilian state of Rio Grande do Sul, Cecília Volkmer-Ribeiro and Maria Christina Mansur, were interested in sponges and molluscs. Before studying these aquatic creatures, they wanted to know more about the quality of the water in different locations on the island. Helped by Sheila Ross from Bristol University and Pedro Mera from INPA, they sampled water from the 'grassy pond', a stream flowing from it, from the lagoons alongside the causeway leading to the station, and from the fast-flowing Uraricoera River. The 'grassy pond' is an unusual example of a pond surrounded by rainforest. It lies a couple of kilometres west of the station, and it is a lovely, mysterious place. The pond itself is about 50 metres in diameter, surrounded by a fringe of graceful *buriti* palms and filled with a profusion of lilies, rushes and other aquatic plants. At the height of the rains the pond is 2 metres deep at its centre, but it dries up during the dry season. The scientists waded boldly into this tangled wetland to conduct their experiments.

Having carefully analysed the waters of these different locations, for acidity (the common pH value, that compares a specimen to a standard solution of potassium hydrogen phthlate), for conductivity and for oxygen, phosphate, silica, iron, nitrate and other chemicals, the team collected sponges and molluscs. They found that the pond was full of sponges, of five species, each of a different genus: every plant or object pulled from the bed of the pond was encrusted with sponges. This abundance is apparently due to high levels of dissolved oxygen and high acidity (which prevented growth of molluscs), while silica content was low since it had been absorbed in the sponge skeletons. Buried in the midst of deep forest, the grassy pond is similar to a peat bog of northern latitudes. Such conditions contain no detectable calcium, so that there were no molluscs in the grassy pond or a nearby stream. Lack of calcium also meant that there was only one species of mollusc in the lagoons or borrow-pits alongside the causeway and three species of bivalves in the main Uraricoera River. The big river has low mineral and pH values – environmental conditions in which molluscs do not thrive.

The basis of the Ecological Survey was of course the botanical inventory of Maracá. All insects and animals found on the island had to be recorded in the context of the surrounding flora, so a good vegetation map was essential. Some 5 per cent of the island was found to be covered in dry savanna and there were patches of wetland, but the majority was *terra firme* upland forest. At the eastern end of the island the forest canopy was fairly low, between 25 and 40 metres above ground, although the occasional 50–metre tall emergent towered above the surrounding trees.

By Amazonian standards, the forests near the station are not very rich in species. Species diversity was however greater in the less accessible forests of the western part of Maracá, where there were familiar Amazonian trees such as *Hevea guianensis*, *Alexa canaracunensis* and abundant Euphorbiaceae. A curious phenomenon is that some parts of the island are dominated by a single tree species, the graceful 40-metre *Peltogyne gracilipes*. This dominance by one type of tree breaks the usual rule for rainforests, whereby many different tree species have to be intermingled to protect themselves from devastation by parasites, insects or tree diseases. Most trees devote much energy to producing fruits, nuts or other mechanisms that will ensure that their seeds are scattered. But Maracá's *Peltogyne* flourish despite being almost a monoculture.

In some parts of the island's forest more than a third of the trees are deciduous, shedding their leaves at different times of the year. This means that those parts should be classified as 'semi-evergreen seasonal forest'. Most of Maracá, however, is covered in conventional evergreen moist forest.

The botanists identified various categories of low forest, including *campina* and *carrasco* thicket forests which occur on sandy soils. Along streams and in some swamps, the vegetation is dominated by lines of grace-

Sunset on the Uraricoera River, a time when biting *pium* black fly retire and thousands of nocturnal bats emerge. (*Chris Caldicott*)

ful *buriti* palms (*Mauritia flexuosa*), each crowned by a halo of fronds. Along the rivers, where the banks are high upland *terra firme*, forest comes right to the water's edge; but where the banks are shallower there are lower communities of *Cecropia* and *Triplaris surinamensis*. If the banks shelve into the river, there are dense tangles of *Inga* shrubs.

The Project's botanical teams collected some 1050 species of higher plants: angiosperms, flowering plants whose seeds are in closed ovaries. This collection includes at least three new species: a legume that has been named *Ormosia maracana*; a Polygonaceae (knot-grass, dock family) which is a new species of *Ruprechtia*; and a bristly, aromatic Boraginaceous herb that is a new species of *Cordia*. Some fifty species of fern were collected, of which five are new to Brazil and twenty-seven are new records for Roraima. Although the project included no fungus team, seventy-five species of fungi were recorded.

In all, the teams of botanists from INPA, Piauí, Kew, Edinburgh and other botanical gardens collected over 2500 numbers of plants. The botanists brought back some 20,000 specimens – many pack-loads and canoes full of great bundles of branches and blossoms. They spent days drying their collections over a variety of cookers, then pressing and recording their plant trophies.

Potentially the most important of all the discoveries of the Maracá Project was a leguminous tree. (Legumes are members of the pea or bean family: they are plants whose seeds grow in pods. There are over 17,500 species of legumes in 670 genera throughout the world, and they are the third largest flowering plant family.) In the months before the Project went into the field, there had been exciting research by the Jodrell Laboratory of the Royal Botanic Gardens, Kew, and by St Mary's Hospital in London. These laboratories had discovered that a tree called *Castanospermum australe*, which grows only in Queensland, contained an alkaloid (castanospermine) that seemed to prevent the dreadful AIDS virus from reproducing in a human cell. Laboratory tests were extremely encouraging. However, the Australian tree is endemic to that region and quite rare. Taxonomists were put to work to try to find other related trees that might contain the same valuable alkaloid. Charles Stirton at Kew suggested that the genus *Alexa*, which grows in Central America and the Guianas and which belongs to the same Sophoreae tribe as *Castanospermum*, might be the answer. Tests on some seeds and leaves in the Kew herbarium confirmed that it was. Therefore, large quantities of *Alexa* were needed in a hurry.

Gwilym Lewis, one of the Project's four excellent botanists from Kew, was told to try at all costs to find *Alexa*. He had worked at Maracá in the early days of the Project and knew that the island's forests are full of leguminous plants. When he returned in October 1987, he was thrilled to learn that William Milliken, a Cambridge graduate and the only botanist permanently resident in the Project, had collected various unusual legumes from distant parts of the island. As Gwil Lewis said: 'I could hardly believe my luck when he flipped open the newspaper protecting the first specimen, to expose a well-preserved example of *Alexa canaracunensis*.' This was precisely the species of *Alexa* that the Kew taxonomists considered most promising as a carrier of castanospermine. William Milliken explained that this specimen had come from the western part of Maracá, 60 kilometres from the base-camp up the treacherous, rapid-infested Maracá Channel to the south of the island.

The two botanists decided to go to collect more *Alexa*, and also to attempt a circumnavigation of the 200 kilometres of rivers that surround the island. Lewis was apprehensive, since he had lost many belongings in an earlier river accident. So the pair travelled light, reckoning to eat fish and forest fruits whenever possible, and they wrapped all their possessions in waterproof containers. They had the great advantage of some Maiongong Indian boatmen, for this tribe are the great navigators of the Uraricoera River and know every one of its countless rapids. Also, it was the start of the dry season, when the rivers were still high and many rapids were covered. In Lewis's words: 'We set off – two tall, thin, pale-skinned Englishmen and three well-muscled, stocky bronzed Indians.' They also had a brilliant Brazilian boatman, Luis Pestana.

> At the more dangerous points he would always motor up to the lip of the rapid, where the water thundered over the protruding boulders, and consider the possible routes before allowing the boat to drift back downstream for 10

metres or so. Without further consideration, he then opened up the throttle and we raced up-river; water crashed over the bows and filled the bottom of the boat and we bailed frantically to keep afloat. Eventually, however, we did have to leave the boat and struggle through the fast currents, dragging the boat over the slimy rocks and broken tree-trunks. Occasionally we had to cut away the tangled riverine forest above our heads or saw through submerged tree-trunks blocking the narrow channels.

We disturbed beautiful birds sunning themselves on the river-banks or on rocks midstream – egrets, herons and diving birds sped away as we approached. Giant Amazonian otters – now an endangered species – were a rare treat, seen bobbing up and down as they sped along in the river currents, calling out to each other like happy children at play. Water snakes, capybara, tapir and alligators were more common, and we were lucky to see pink river dolphins.

After two days, the botanists had reached the far western tip of Maracá and they started to return down the northern Santa Rosa Channel. They had to pass the mighty Purumame Falls. This involved carrying all their supplies up and over the portage trail, which ran for a couple of kilometres over a steep granite hill in the forest beside the thundering falls. They also had to make a sled to drag their metal boat; it hurtled down the near-vertical muddy path on the eastern side of the hill. One tough young Maiongong hefted the very heavy 25-hp Mercury outboard motor on to his shoulders and carried it for the entire distance – a feat that almost no other member of the Project could have accomplished.

At the base of the portage is an agreeable sandy valley, a perfect camp-site close to the churning waters of the Falls. Famous earlier explorers have carved their names on rock outcrops here: Schomburgk drew an elegant 'S' in 1838; Alexander Hamilton-Rice in 1925 left a message and an engraving of his seaplane, the first aircraft in Roraima; and in 1927 the great Brazilian explorer and champion of the Indians, Cândido Rondon, left his name.

Lewis and Milliken set off to collect near the camp, while the Indians fished.

Within twenty minutes we had found a specimen of *Alexa canaracunensis* and, as previously confirmed by William, the cut bark gave off a strong odour of freshly sliced cucumber. The tree-trunk was about 10 centimetres in diameter, with fissured greyish-white bark and an elegant crown of large pinnate leaves.

William Milliken climbed the tree to cut flower-heads and leaves from its crown, but he soon slid down in a hurry, having been bitten by little black ants that live in the hollow stems of the branchlets. This was a new discovery, another example of symbiosis. Tiny glands on each flower secrete a sugary food to attract the ants, who in turn protect the host tree.

The two botanists and their Indian helpers collected plenty of fruit, bark, resin and leaves of the *Alexa*, particularly along a trail that we had cut across steep hills at the western end of Maracá. They also gathered material from many other leguminous plants, including a tree, *Clathrotropis macrocarpa*, related to *Alexa* and giving off the same cucumber smell when slashed. Back at Kew, the specimens collected on Maracá proved to be rich in castanospermine. St Mary's Hospital in London is clinically testing this drug on human beings. If it were to prove useful in combating AIDS, this would be another potent argument for protecting rainforests. With the untold diversity of genetic resources in those forests, there is no knowing how many other potentially valuable species they contain.

In many parts of the tropics, forests end in a steep wall of vegetation beyond which is open grassy savanna studded with a few short, gnarled trees. Why should this abrupt change occur? What causes this violent transition between two totally different ecosystems? The two competing types of vegetation enjoy the same climate, with identical rainfall and tropical sun, and they are essentially at the same altitude. Scientists have often puzzled over this mystery.

One of Maracá's many attractions was its location on a biogeographical boundary between forest and savanna. The island itself contains both types of terrain. To the west are hundreds of kilometres of unbroken tropical forest; to the east are the open plains of the upper Rio Branco.

John Proctor's Forest Regeneration team joined soil scientists, botanists and entomologists from other programmes to try to solve this problem. Trenches and pits were dug across forest–savanna boundaries to look for changes in soil types. One such trench was close to the low plateau on which the ecological station is built; the other was across a large grassy clearing a few kilometres away at Santa Rosa on the northern edge of the island.

Soil chemistry did vary across the transects. The savanna areas had more acid soil. Where the ground rises slightly as it enters the forest, 'exchangeable cations' (calcium, magnesium and to a lesser extent potassium) were more concentrated, and the soil had a higher pH factor. Not surprisingly, there was also a higher content of organic carbon in soils beneath the trees, owing to the mass of leaf litter on the forest floor. But none of this was sufficient to explain the dramatic change in the overlying vegetation. The variations in soils were probably a result of the forest–savanna boundary, rather than its cause.

In another experiment, measurements were made of the breakdown of soil organic nitrogen – its mineralisation and subsequent nitrification. These are key processes in most ecosystems, and are particularly vigorous in tropical forests. Rob Marrs of the Institute of Terrestrial Ecology found that mineralisation and nitrification of soils varied between the wet and dry seasons at Maracá, but that rates were lower in forests where there was little seasonal change. An important discovery was that these essential processes were faster at times of weather transition, during periods of wetting and drying. Mineralisation and nitrification were always less in soils under savanna than under forest. Another significant finding was that nitrogen mineralisation was immobilised under waterlogged savanna. Even nutrients added experimentally to flooded soils were immobilised rather than nitrified. This may be an important nutrient-conservation mechanism.

Perhaps water-level was the explanation for the abrupt change of ecosystem from forest to savanna. Monitoring the height of ground water in these pits showed that water-table levels were probably the single most important factor influencing the change from forest to savanna vegetation. At the height of the rainy season, the water-table reached the surface of the open areas and lapped at the roots of trees in the surrounding forest. Soils beneath the grassy clearings were saturated during the three months of high rainfall, whereas those beneath nearby forest were not.

Permanent plots were established across the forest–savanna boundary to monitor vegetation change. The scientists wanted to see whether either type of competing vegetation was advancing. It became apparent that on Maracá, where there is no human interference, forest is invading open grassy areas (which are known in Brazil as *campo*). Occasional *campo* trees such as the typical gnarled *Curatella americana* were found inside the forest, near its edge. They had clearly been overwhelmed by the advancing rainforest,and would not long survive being shaded by the higher canopy.

The forest margin has a distinct vegetation. Some species are typical of edge-sites: trees whose branches overhang the grassy *campo*, small palms and shrubs, and a marginal thicket of vines – all plants taking advantage of the sunlight. This vegetation gives the forest edge its wall-like appearance. It protects the more open forest beyond from lateral winds. The marginal plants may also provide the shade and the higher levels of organic matter needed to establish forest vegetation at the savanna edge. This could be an important lesson for those seeking to reafforest lands denuded of their rainforest cover. Monitoring of the permanent plots during coming years should show whether the forest is still advancing.

Apart from studying the two types of flora, the rise and fall of the fluctuating water-table, and changes in soil chemistry and composition, the Project's scientists also examined soil micro-organisms beneath the forest–savanna transition. Lucille Antony, the head of INPA's ecology department, led this research. She found that soil micro-organisms congregated more densely below the actual boundary than below the forest or *campo* on either side. These tiny organisms account for a large share of the total non-plant biomass in a rainforest. They include a great many of the thousands of species yet to be discovered. But here on Maracá they appeared not to be the cause of forest–savanna change; as with other factors, they were the product rather than the cause of the boundary.

In the interior of Maracá Island there are some clearings covered in a different type of vegetation. This is *vazante*, a dense tangle of tall herbs and low shrubs. *Vazante* is almost treeless, but tightly packed with grasses, herbaceous vines and Convolvulaceae. It has been aptly compared to an abandoned and overgrown British allotment site (see page 53). Many plants in vazantes are semi-aquatic, apparently because these areas lie along drainage lines – they do not have permanent streams, but suffer flooding during the rainy season. *Vazante* soils are dark clays which are clearly good reservoirs of moisture and nutrients. This may account for the contrast between their thick, low vegetation and the open grassland of conventional *campo*.

The forest surrounding *vazante* clearings ends in abrupt, steep walls. One such area showed up on satellite imagery as a neat rectangle covering some 10 hectares. With its straight sides and rectangular corners, it looked man-made. This was an intriguing mystery, as we were sure that no one lived on Maracá Island. We therefore mounted a small expedition to investigate. This involved a trip up the river to the north of the island and a short cut inland through riverine forest. But when we reached the *vazante* rectangle it proved to be natural: on the ground, the surrounding forest walls were not quite as straight and rectangular as they had appeared in satellite pixels.

John Proctor's team found that some forests occurring at the eastern end of Maracá Island are unusual. They are different from forests in other parts of the tropics and different even from forests farther south in the heart of the Amazon basin. They have a low density of trees and relatively small number of tree species. With a long dry season and fairly low annual rainfall, they are technically tropical moist forests. The system is, nevertheless, evergreen – with only 6 per cent of its trees deciduous or leaf-shedding. However, most of the forests at the eastern end of the island are very different. Here, Jim Ratter and William Milliken found fairly typical tall Amazonian forest, much of it classified as 'exceptionally large forest' on the criterion of having a total trunk basal area of larger trees exceeding 40 square metres per hectare.

This forest grows on the weakest and sandiest soil that the scientists had ever seen under any rainforest. But Maracá's trees do not react to such infertile soil in the expected way. They have a relatively low mass of roots, whereas it was thought that vegetation on poor soil had to have a dense root mat on the soil surface to ensure efficient nutrient cycling. Instead of spreading their roots horizontally over the surface to collect all possible nutrients from the decomposing litter, Maracá's trees paradoxically put roots down into the weak soil. Trees on such soils were thought to contain little phosphorus, but dead leaves on Maracá have high concentrations of this element. Such sandy soil would usually be expected to support heath rather than tropical forest, whereas Maracá's forests are normal in appearance.

Another interesting aspect of this forest is the frequency of buttress roots and epiphytes. Many trees in Amazonia have thin, plate-like buttress roots projecting from the base of their trunks. The purpose of these buttresses is apparently to hold the tree upright (see page 43), and about 12 per cent of Maracá's have such buttresses. Only 15 per cent have epiphytes, the perching plants that grow in tree branches but are not parasitic in feeding off their hosts. This low percentage is thought to be a result of Maracá's long dry season. On the other hand, lianas or creepers are very common. Walking through these woods, you often come across dramatic baroque tangles of creepers that form part of everyone's vision of a tropical jungle.

John Proctor found a layer of charcoal or ash under some of the forest at the eastern end of the island. He wondered whether the peculiarities of this forest might be partly explained by its being secondary growth that had once been disturbed by man. It is true that there were native Indians living on or near the island in recent centuries. When the German anthropologist Theodor Koch-Grünberg was there in 1911, he visited two small families of Indians living near Maracá's eastern tip. Years later, when the island became a protected reserve in the 1970s, a family of settlers was evicted from a hut on the Santa Rosa Channel. But these references to habitation on Maracá are too few to have caused human deforestation there. Ever since cattle were introduced to Roraima two centuries ago, there has been ample open savanna for them. Settlers have never needed to clear forest until very recently, and we know that Maracá has not been deforested in recent times. Nor would Indians have made large-scale felling. They make only small slash-and-burn clearings which are soon abandoned for reoccupation by the surrounding forest. Indians generally prefer forest rich in game to open grasslands containing only a few deer. So Indians would not have had the incentive to deforest eastern Maracá, nor the tools to do so.

John Proctor's Forest Regeneration team – Jill

Thompson, Duncan Scott and various helpers – spent two full years of hard work at the University of Stirling in Scotland. They were analysing the mass of soil and leaf samples and other data they collected during thirteen months on Maracá. The Forest Regeneration programme will involve regular return visits to the island to monitor vegetation changes in the artificial clearings. This research has been funded by a series of grants from the Overseas Development Administration, which sees the research as important for the regeneration of all tropical forests. But the analysis of such a vast amount of information has inevitably been slow.

The Forest Regeneration team went back to Maracá in 1989 to observe changes in its series of artificial gaps. It found that regeneration was rapid and vigorous. Pioneer trees, tree seedlings and regrowth from cut stumps all contributed to the mass of new vegetation. A layer of herbs grew quickly in the clearings, covering the ground with shade needed by saplings of big shade-seeking forest trees. This herb layer increased greatly in height and vigour during the first few years; but it was a temporary phenomenon that will disappear as new trees grow. This is a useful discovery. Reforestation of denuded rainforests can be helped by establishing a layer of herbs under whose shadow larger trees may start to grow.

Much future research on rainforests will be concerned with their underlying soils. It is normal for tropical forests to grow on very weak soils. With their high rainfall and temperatures, such forests photosynthesise intensively and hence produce luxuriant, fast-growing vegetation. This vigorous vegetation recaptures most nutrients before they enter the soils beneath, and it can remove such nutrients as there are in the soils. The soils themselves tend to have suffered intense weathering, so that they have lost unstable base ions and retain more stable minerals such as kaolinite, quartz, iron and aluminium oxide. In the Maracá Project, study of the soils was fundamental to various research programmes: to the main Ecological Survey, to the Soils and Hydrology programme, the Forest Regeneration programme and the study of the forest–savanna boundary.

Two tough and energetic soil scientists, Dan Robison and Stephen Nortcliff, took teams of woodsmen and technicians into the remotest unexplored parts of the island to dig soil pits. Their work was another example of scientists quietly performing tasks that publicity-seeking 'explorers' would have trumpeted as heroic feats. At the Royal Geographical Society we are pestered by some of these bogus explorers, who boast about their solo exploits and get plenty of media attention. Many of us, and particularly the soil scientists and botanists, cut trails into tough terrain knowing for sure that no so-called civilised man and probably no Indian had ever set foot there. This was hot, exhausting and sweaty work, mildly dangerous and rewarded by many bites and scratches. But it was true exploration in a worthwhile cause.

Robison and Nortcliff made 144 observations and dug thirty-one pit profiles in different parts of the forested Reserve. They hauled back soil samples which were then tested in many ways: for the exchangeable cations (calcium, magnesium and sodium), for acidity, cation exchange capacity, for pH, nitrogen, phosphorus, iron oxides, aluminium oxides and calcium and magnesium using ammonium acetate as the exchange extractant. They also made minute observations of the texture, soil type, colour, grain and porosity of their samples.

These tests showed that the soils of Maracá were far more diverse and complicated than had been imagined. The underlying geology of this part of Brazil is the pre-Cambrian Guiana Shield – a vast formation of very ancient sandstones and granite that underlies all the northern extremity of South America. These are some of the oldest rocks to be seen anywhere on our planet: they were formed millions of years before continental drift caused the South American continent to move gradually westwards away from Africa. At Maracá there are also ancient sedimentary rocks deposited on the original Guiana Shield material. These sediments are known as 'drift', as opposed to the 'residual' bedrock.

The foot-slogging groundwork of the pedologists (the technical name for soil scientists) showed that Maracá Island can be divided into five main physiographic regions. The flat area near the ecological station at the eastern end of the island has sandy soil lying on a lateritic conglomerate of pebbles cemented by iron and quartz. This is where most of the Project's research took place. The scientific name for these poor soils is Ultisol, or yellow or red–yellow podzolics. It is remarkable that a rainforest can flourish on such shoddy soils. They are acidic and, depending on the type of 'drift' formation on which they lie, are either very sandy or clayey. Varieties of these soils occurred on the low-lying seasonally flooded stream catchment in the east-central part of Maracá and on the nearby undulating hills of residual granite along what we called the Fumaça trail.

Surprises came in the central highlands, where hills rise to a few hundred metres above the Uraricoera River. Here the pedologists were amazed to find quite different soils on adjacent hills. The geology of the hills themselves varied, some with metamorphic rocks and others with remnants of earlier land surfaces composed of extremely weathered 'drift'. The soils on the metamorphic hills are eutrophic yellow podzolics (or Alfisols), whereas the other hills have red latosols (Ustoxes) which are some of the richest soils under any rainforests. At the western end of Maracá, the range of outcrops that cause the spectacular Purumame waterfall is composed of schists. The soils there are, once again, weak yellow podzols or dystrophic Cambissols – soil types also known as Udults and Dystropepts.

All this painstaking research revealed an extremely wide range of soils on Maracá. The differences are partly the result of changes in the underlying geology. But they show yet again the extraordinary complexity of any tropical forest ecosystem. It was found, for instance, that latosols on level ground can support forest with a higher canopy than can the white highly leached podzols along the sloping rims of valleys.

There are two cycles operating in every tropical forest: the nutrient cycle (concerned with the recovery of nutrients from dead matter) and the hydrological or water cycle. The vigorous vegetation recaptures almost all nutrients from fallen litter and most of the rainfall that drops from leaves or pours down tree-trunks. These two cycles are the food and drink of a living forest.

In the second Chapter, I described the experiments set up by the Soils and Hydrology team led by Professor John Thornes and his colleague Dr Sheila Ross. They created a series of imaginative experiments, and they worked extremely hard in gathering samples and analysing them. It was all done to demonstrate how forest soils react.

The experiments of the Soils and Hydrology team confirmed the obvious finding that more rainfall hits the ground when the forest canopy has been removed – since no rainwater is held by the leaves and later evaporated, and none runs down tree-trunks. The moisture content of soils was shown to relate partly to the location on a slope: sites on mid-slope and at the foot of a slope remain wetter because their soils contain more clay and ferralitic material. Differences in texture also have a bearing on soil moisture tension. Using sophisticated data collected from various locations, at different depths and in different weather, the team devised a computerised model of water through-flow beneath forest soils in relation to rainfall. It was further proved that the volume of water that runs off the surface is controlled by both the slope and by the degree of vegetation clearance. More water runs off towards the base of a slope than at its top. Removal of the root-and-litter mat increases both the rate of surface run-off and the depth to which water penetrates the underlying soil.

The importance of the roots and vegetable litter was demonstrated in the erosion experiments. Raindrops hit such exposed soils with full force, unchecked by the tree foliage. Unprotected soil particles splash up and roll down slopes. Thus, very large losses of mineral soil were observed on mid-slopes under totally cleared forest. The same rainstorms hitting undisturbed virgin forest remove much leaf litter but almost no mineral soil material. The situation is similar under areas where the canopy but not the lower vegetation had been removed. It was clear that the continual shedding of leaves by evergreen forest protects the soil below.

The floor of any forest is a carpet of leaves in varying states of decomposition. This leaf litter creates a humid environment that prevents water evaporating from the soil and filters rainwater into the ground. It also provides the stock of organic nutrients from which the trees' roots draw their nourishment, either directly or through the soil. To discover the quantity of dead litter, the Soils and Hydrology team made monthly collections and weighed and sieved what they had gathered. The samples were analysed for organic and chemical compounds.

These experiments dramatically confirmed the importance of the root-and-litter layer. When it remained intact, erosion was minimal, even when 70 per cent of the overhead vegetation was removed. But when the ground was cleared of all cover, erosion was very bad:

sediment yield increased sharply, to 9 tons of sediment per hectare per year.

Another experiment by this team involved nylon-mesh bags filled with clean leaves of two common trees, *Protium* and *Ecclinusa*. Bag samples were collected after thirty, sixty and ninety days. The bags had small holes below, but 4-millimetre holes above, so that termites could easily enter to start the break-up of leaves and the decomposition process. It was found that the leaves of *Protium* decayed faster than those of *Ecclinusa*. Decay under virgin forest was faster than in the cleared plots, possibly because the termites and other decaying organisms worked better in the shade and humidity of the undisturbed forest.

Maracá's location on a natural boundary between undisturbed forest and open plains yielded other research benefits. It meant that the island is close to a frontier of human settlement – one of the places where men are still expanding by colonising virgin land. They are clearing the edges of the rainforests to try to create more cattle pasture. This made it a perfect place for human-geography studies, away from the protected and uninhabited ecological reserve.

One of the Project's five research programmes was a study of this frontier settlement. We called this the Land Development programme, and it was led by Peter Furley, a geographer from the University of Edinburgh. His team included a number of researchers of different disciplines doing a diverse range of studies, but the sum of their work combined surprisingly well, to form an interesting picture of human occupation of this northernmost part of Brazil.

I set the scene in Chapter 1 with historical studies of the way in which Roraima came to be part of Brazil. In 1943, the Vargas government of Brazil separated what is now Roraima from the state of Amazonas. This frontier region was designated as the Federal Territory of Rio Branco, in order to stimulate its colonisation and thus protect Brazil's remote and sparsely inhabited borders. This administrative separation led to an immediate influx of civil servants and their families, and there were attempts to attract settlers into new agricultural colonies. The first governor of the new Territory engaged a firm of town planners to lay out the streets of its capital, Boa Vista, in their present spacious spider's-web pattern radiating from a central administrative hub.

Roraima was always held back by its isolation. In the late nineteenth century, Amazonia experienced a boom from its monopoly of the world's rubber. State governments in Manaus were desperate for meat from Rio Branco's cattle, but boats loaded with small consignments of animals took three months on the gruelling round trip along the Rio Branco and the Rio Negro. A trail was cut for hundred of kilometres from Manaus to Boa Vista in the 1890s, but it did not solve Roraima's isolation. The advent of air travel in the 1960s helped. But it was not until 1977 that an 850-kilometre unpaved highway was finally opened, after six years of tough labour and many deaths among the construction crews. Roraima's population doubled in the next decade, to 120,000, and the city of Boa Vista expanded in line with the rest of the Territory. The region achieved statehood as Roraima in 1988.

Ever since the 1930s, there has been sporadic prospecting of gold and diamonds in Roraima's remote forested rivers. In 1988 this turned into a major gold-rush, with the discovery of significant deposits of gold on the upper Mucajaí, Uraricoera and Uraricaá rivers. An estimated 40,000 *garimpeiro* gold prospectors flooded into the newly created state. This inrush of adventurers increased the state's population by a third and strained its fragile economy and environment. The tragedy was compounded by the fact that the main gold strike lay within the territory of the Yanomami Indians, the largest surviving forest tribe in South America.

The 17,000 Yanomami live in thatched *malocas* amid the densely forested Parima hills of western Roraima and on the headwaters of the Orinoco in Venezuela. The tribe has been isolated until recent times, so that it was unprepared for the shock of contact with the mining prospectors. The *garimpeiros* unwittingly brought new diseases, notably virulent malaria, that have killed hundreds of Indians; their tough attitudes undermine tribal society and culture; their flights into dozens of clandestine jungle airstrips frighten off the game hunted by the natives; they destroy forests and riverbeds with their mining operations; and the mercury they use to refine gold poisons river waters and fish.

Almost half Roraima's territory is protected, either as Indian reserves or as uninhabited forests set aside for conservation. The 110,000-hectare Maracá Ecological Reserve is one of Roraima's smaller protected areas. By far the largest protected area is the national park for the Yanomami. Pro-Indian activists long campaigned for the western part of Roraima and northern edge of Amazonas states to be reserved as a homeland for these 8000 people in Brazil, who are its only inhabitants. In 1982, an area of 7.7 million hectares was 'interdicted' for the Yanomami; but it has never been legally enacted and it is seriously threatened. The government of President Fernando Collor de Mello tried to evict the thousands of prospectors who had illegally invaded the area. Then, in 1992, it finally legalised a huge protected territory for the Yanomami Indian nation.

The natural open savannas of Roraima have long been occupied by plains Indians. Most have been wiped out – by disease, by social disruption with the occupation of their lands by cattle ranches, and by forced exile in the late eighteenth century. The main surviving groups are the Carib-speaking Makuxi and the Aruak-speaking Wapixana. These are now almost fully acculturated into frontier society. They live in protected villages surrounded by cattle ranches. By the first half of this century, the total population of Roraima's plains Indians had fallen to some 6000. Their numbers have increased dramatically, tripling during the past thirty years to some 18,000. This welcome rise is due to the efforts of missionaries and the Indian service, to social stability resulting from greater legal protection, and to improved medical and natal services. All Indians are, however, subject to constant invasion of their remaining pockets of land and to occasional physical violence.

The new state of Roraima has the highest ratio of Indians in its population of any part of Brazil. It also enjoys the lowest density – 0.34 inhabitants per square kilometre compared to 14.07 for all Brazil – and a very youthful demographic profile. Even before the influx of gold prospectors, Roraima's population was growing faster than most of the rest of the country. Inward migration accelerated with the opening of the Manaus–Boa Vista road in 1977. It is now fairly easy for settlers from other parts of Brazil to bring their families and animals to this frontier region.

A team of human-geographers from Belém in Pará studied these colonists, at three locations not far from Maracá Island. The principal investigators were Luc Mougeot, a Canadian professor at the University of Pará, the French researcher Philippe Léna from the Goeldi Museum in Belém, and the Brazilian economist Nelson Vieira. They found from their many detailed interviews with settlers that the main factor in the success of an agricultural colony lay in the transport of its produce to market in Boa Vista. The most remote area at Tepequém was failing because of the length and roughness of its road link to the state capital. By contrast, the relatively old colony of Taiano (founded in 1953) was prospering because of its proximity to Boa Vista. It also enjoyed a small patch of fertile soil, beneath what had been semi-deciduous forest. Such seasonal trees grow on better soil than does pure tropical rainforest. However, where those same colonists tried to farm marginal land at the edge of rainforest they achieved only short-term gain but caused long-term environmental degradation.

Tom Dargie, an independent consultant formerly with Sheffield University, and Peter Furley of the University of Edinburgh sought to measure the rate of deforestation and land-use change along Roraima's northwestern settlement frontier. They compared different attempts at remote sensing of this region during the past twenty years. In the 1970s there was a survey by Projeto Radam, using SLAR (side-looking airborne radar) at a resolution in units of only 1 square kilometre. Then came satellite monitoring by NOAA, by the French SPOT and by Landsat. These were on far more detailed resolutions, at scales of 1:20 square metres and then 1:10 square metres. The different satellites use different systems and produce images at distinct wavebands, so that they may be used for different interpretations.

This research team returned to Edinburgh with its mass of field 'ground-truth' data and then developed a computer program to merge all the available surveys into a database for Maracá and its surroundings. They developed a GIS (geographic information system), which is the transfer of geographical and cartographic information into digitised form that a computer can read and manipulate. Many hours of time on a powerful computer were used to produce this GIS and resulting maps of Maracá's ecology.

We purchased Landsat imagery and were able to compare changes using data from 1978 and 1985. Tom Dargie's painfully acquired ground-truth fixes were essential to interpret the imagery. The findings are disturbing. Even in this remote part of Brazil, at a time when settlement was considered minimal, deforestation is proceeding at an alarming rate. During those seven years, 4 per cent of the study area was deforested. To the east are vulnerable and extremely important remnant forests and seasonally flooded gallery forest and *várzea*. The team found that almost 25 per cent of these was deforested in those seven years. The negative result of that clearance emerged in research by Chris Barrow of the University of Wales, but he also showed that commercial rice production is flourishing on flooded *várzeas* alongside rivers near Boa Vista. As a result, the state is self-sufficient in this food crop.

Mike Eden and Duncan McGregor of the University of London studied the effect on soils of all this forest clearance. They asked farmers when they had deforested each field, and studied what had happened to areas cleared from one to fifteen years previously. They found that the soils beneath such areas change rapidly and deteriorate disastrously. Topsoil becomes denser, compacted by rainfall and trampling by cattle. Immediately after clearance the topsoil gains from nutrients released by burned vegetation. But the beneficial effects are very short-lived. The soil's pH increases from burning, but then declines steadily. Organic carbon is never high in the weak soils beneath tropical forest. The same is true of phosphorus, which increases from burning but still remains below the level needed for good pasture; it then declines quickly, rendering the soils sadly infertile.

This research dramatically confirmed that tropical rainforest cannot be converted into adequate cattle pasture, except in a few areas of exceptional soils. In Roraima, land cleared of forest could support only a few animals – too few for such ranches to be commercially viable. Our researchers also studied a well-capitalised ranch; but they found that its higher density of cattle accelerated the soils' deterioration, despite attempts to apply fertilisers. Cattle are imported and inappropriate animals for tropical forest regions. This harsh lesson is being learned by all levels of society, by legislators including Brazil's dynamic young President Fernando Collor de Mello, by big landowners and the banks that finance them, and by poor settlers. The lesson cannot be learned too quickly, before more magnificent rainforests are destroyed in this futile practice. President Collor declared, in a visit to the Royal Geographical Society, that he was determined to put a stop to forest clearance to create cattle pasture, because 'It is pointless waste, from which nobody gains.'

WHAT was achieved by the Maracá Rainforest Project? The original invitation from Dr Paulo Nogueira Neto was to make an Ecological Survey of the Reserve. He also wanted zonation of the island's ecosystems and a plan for future research. Obviously, we could not hope to study more than a fraction of the thousands of species of flora and fauna in that rich rainforest; but we did fulfil the general aim.

The five research programmes that were undertaken in addition to the Ecological Survey all broke new ground. They added significantly to understanding how rainforests function, how they regenerate under different types of destruction, and how rainfall and soils interact. The Medical Entomology programme contributed to a larger campaign to control insect vectors that spread so much misery in the tropics. And the Land Development programme made valuable studies of a settlement frontier, just as Roraima was starting to expand explosively from an influx of *garimpeiro* gold prospectors. It also contributed to knowledge of how to interpret satellite imagery in order to assess the rate of rainforest felling.

Another achievement of the Project was its size. It was one of the largest multi-disciplinary expeditions ever mounted in Amazonia. Many Brazilian scientists were suspicious of this size at the outset, before we started the field phase, but eventually word spread that this was a good place for research, and excellent Brazilians poured in. Maracá was served by a competent and hard-working administrative team, so that supplies flowed regularly to the camp, and it had medical backup. The presence of so many leading scientists came to be seen as an asset – there was fruitful interchange between scientists of different disciplines on the station's veranda at night. So, as the months went by, we received requests from all over Brazil from people wanting to join. In the end, four times as many Brazilian scientists as British participated, although, as mentioned earlier, the smaller British contingent tended to stay for longer periods, so that it worked for roughly the same number of man-days.

The Project received a good press in Brazil. Various national newspapers and magazines covered it extensively. It was also featured in the *Fantástico* programme on Globo television, a very popular Sunday evening magazine programme with an audience of tens of millions.

We were determined to try to present the results first in Brazil and preferably in Portuguese. There was a workshop in Boa Vista at the close of the field phase in March 1988. Many scientists assembled in Manaus in April 1989 for the first review conference and a second review conference was held at the Royal Geographical Society in London that September, and the Project arranged for a dozen Brazilian scientists to attend. Many of the scientific papers are being published in special Maracá supplements of INPA's journal, *Acta Amazonica*.

Various factors meant that the Brazilian authorities warmly welcomed the Project. It had gone to Brazil at the invitation of the Environment Secretariat, to do work wanted by the host country. It undertook non-political, uncontroversial research. It involved a preponderance of scientists and technicians from Brazil, and the foreign members tended to be people who already knew (and loved) Brazil and spoke Portuguese.

Thus, when the British Foreign Office Minister Tim Eggar visited Brazil, he found that President José Sarney warmly praised the Maracá Project. It was also cited as a fine example of international scientific cooperation by the next President of Brazil, Fernando Collor de Mello; by Fernando Mesquita, the first President of the new environmental agency IBAMA (which incorporated our original host, SEMA); and by the new Brazilian Environment Secretary, José Lutzenberger. This led directly to the signing of a protocol for cooperation in rainforest research between Britain and Brazil. It was the first such inter-governmental agreement that Brazil had ever signed for such research. The British Minister for Overseas Development, Chris Patten, signed the accord in Brasilia in June 1989, a few months after the second Maracá review conference.

In a speech to a conference at the RGS in 1990, His Royal Highness The Prince of Wales summed up the Project's success. He said:

> The Maracá Rainforest Project (of which I was privileged to be the Patron) made the most intensive study ever undertaken of a forest at the Northern edge of the Amazon Basin. It has already identified some 250 species new to science, including an *Alexa* tree that is rich in castanospermine, a plant alkaloid that may prove to be of great benefit to humankind in its fight against AIDS. It is impossible to over-estimate the importance of research and education. We all have so much to learn.
>
> Of equal importance are those direct links between the countries involved in projects such as these. The Maracá Rainforest Project was organised by the Royal Geographical Society at the invitation of the Brazilian Environment Secretariat.
>
> I think I can safely claim that the Maracá Project was well received by the Brazilian authorities, for many different reasons. It involved thirty British scientists working alongside 120 Brazilian colleagues, and a further fifty Brazilian scientific technicians. Most of the British contingent knew (and loved) Brazil, so that all but two of them were able to present their papers at the first Review Conference, in Manaus, in Portuguese. The five research programmes which made up the project were all carrying out work requested by the Brazilians themselves, and the results are being published where it matters most: in Brazil.
>
> The relationship was cemented when Britain and Brazil signed an inter-government agreement for environmental cooperation – the only such agreement that Brazil has so far undertaken with any country. At a time when North and South find it hard to communicate on these issues, I set great store on such practical initiatives.

APPENDIX ONE: MEMBERSHIP OF THE MARACÁ RAINFOREST PROJECT

1. *Ecological Survey*

INPA = Instituto Nacional de Pesquisas da Amazônia
(National Amazon Research Institute in Manaus)

BOTANY

Dr James Ratter (botanist/ecologist)	Royal Botanic Garden, Edinburgh
Dr Ângelo Augusto dos Santos (ecologist)	INPA-Ecologia
Dra Maria Lúcia Absy (palynologist)	INPA-Botânica
Dra Maria das Graças Medina Arrais (bromeliads)	University of Piauí
Peter Edwards (ferns)	Royal Botanic Gardens, Kew
Carlos Alberto Alves Freitas (ferns)	INPA-Botânica
Dr Raymond Harley (Labiatae herbs)	Royal Botanic Gardens, Kew
Dr Mike Hopkins (ecologist)	University of Papua New Guinea
Dra Isonete de Jesus (fungi)	INPA-Botânica
Gwilym Lewis (legumes)	Royal Botanic Gardens, Kew
Pedro Augusto Suárez Mera (algologist/limnologist)	INPA-Limnologia
William Milliken (ecologist)	Royal Geographical Society
Dr Marçal Queiroz Paulo (lichens)	University of Paraíba
Dra Marlene Freitas da Silva (angiosperms)	INPA-Botânica
Edileuza Sette Silva (lichens)	Museu Integrado, Roraima
Cynthia Sothers (ecologist)	Worldwide Fund for Nature – Manaus
Luis Augusto Gomes de Souza (legumes)	INPA-Agronomia
Brian Stannard (parasitic Loranthaceae)	Royal Botanic Gardens, Kew
Prof. Lauro Xavier Filho (lichens)	University of Paraíba

ENTOMOLOGY

Nair Otaviano Aguiar (Coleoptera)	INPA/University of Amazonas
Luis Fernando Reys Aguiar (dragonflies)	Museu Nacional, Rio de Janeiro
Adelmar Gomes Bandeira (termites)	INPA
Dr Forbes Benton (ants, Phoridae)	CEPLAC (Cocoa Research Centre), Itabuna, Bahia
Erica Helena Buckup (spiders)	Fundação Zoobotânica do Rio Grande do Sul
José Ramiro Botelho (ectoparasites)	University of Minas Gerais, Belo Horizonte
Carlos Roberto Brandão (ants)	University of São Paulo
Lúcio Antônio de Oliveira Campos (bees)	University of Viçosa, Minas Gerais
Eliana Maria Cancello (termites)	University of São Paulo, Zoologia
Cláudio José Barros de Carvalho (Diptera)	University of Paraná, Curitiba
Dra Mirna Martins Casagrande (butterflies)	University of Paraná, Curitiba
Márcia Souto Couri	Museu Nacional, Rio de Janeiro
Ana Celeste Ferreira (Plecoptera)	INPA-Entomologia
Maria Helena Galileo (Coleoptera)	Fundação Zoobotânica do Rio Grande do Sul
Marcos Vinicius Bastos Garcia (bees)	WWF, Manaus
Sandra Harrison	Museu Nacional, Rio de Janeiro
Dr Pierre Jolivet (beetles, ant plants)	University of Viçosa, Minas Gerais
Dr Pedro Marcos Linardi (ectoparasites)	University of Minas Gerais, Belo Horizonte
Prof. Arno Antônio Lise (spiders)	Museu de Ciências Naturais, Porto Alegre, Rio Grande do Sul
Paulo Machado (ectoparasites)	University of Minas Gerais, Belo Horizonte
Maria Aparecida de Leão Marques (spiders)	Fundaçâo Zoobotânica do Rio Grande do Sul
Hugo Mesquita (Odonata)	INPA-Entomologia
Dr Olaf Hermann Hendrik Mielke (butterflies)	University of Paraná, Curitiba
Padre Jesús Moure (bees)	University of Paraná, Curitiba
Francisco Peralta (bees)	INPA-Entomologia
Dr José Albertino Rafael (horseflies)	INPA-Entomologia
Dr Anthony Raw (wasps)	University of Brasília
Prof. Gilberto Righi (annelids)	University of São Paulo
Raquel Telles M. Sampaio (aquatic insects)	INPA-Entomologia
Jaime Maia dos Santos (nematodes)	University of Viçosa, Minas Gerais
Rosa Maria do Socorro dos Santos	INPA-Entomologia
Manoel do Nascimento Silva	Department of Education, Roraima
Og Francisco Fonseca de Souza (termites)	University of Viçosa, Minas Gerais
Maria Elizabeth L. de Souza (Hemiptera)	Fundação Zoobotânica do Rio Grande do Sul
Dr Ivaldo Ferreira Vilela (ants)	University of Viçosa, Minas Gerais

ZOOLOGY

Dr José Márcio Ayres (primates)	Museu Paraense Emílio Goeldi, Belém
Adrian Barnett (small mammals)	Independent
Prof. Roberto Cavalcante (birds)	University of Brasília, Ecologia
Mário Cohn-Haft (birds)	INPA/WWF Minimum Critical Size of Ecosystems
Elton Pinto Colares (giant otters)	INPA-Biologia de Água Doce
Aléxia Celeste da Cunha (small mammals)	INPA
Efrem Jorge Ferreira (fish)	INPA-Biologia de Água Doce
José Manuel Fragoso (peccaries)	University of Florida
Natália Inagaki (primates)	Museu Paraense Emílio Goeldi, Belém
Michelle Jegu (fish)	INPA
Olavo Lira (fish)	INPA
Dra Maria Cristina Dreher Mansur (molluscs)	Fundação Zoobotânica do Rio Grande do Sul
Márcio Martins (frogs)	INPA-Ecologia
Rogério Gribel Soares Neto (bats)	INPA-Ecologia
Andrea Portela Nunes (primates)	University of Rio de Janeiro/INPA
Mark O'Shea (reptiles)	Independent
Sharon Pollard (bats)	WWF/INPA-Ecologia
Paulo Petry (fish)	INPA-Biologia de Água Doce
Aldenora Lima de Queiroz (mammals)	INPA-Ecologia
George Rebelo (caimans)	INPA-Ecologia
Profa. Cecília Volkmer Ribeiro (sponges)	Fundação Zoobotânica do Rio Grande do Sul
Barbara Ann Robertson (zooplankton)	INPA-Limnologia
Fif Robinson (bats)	University of Oxford
José Maria Cardosa da Silva (birds)	Museu Paraense Emílio Goeldi, Belém
Maria Nazareth Ferreira da Silva (bats)	INPA-Ecologia
Douglas Stolz (birds)	INPA
Aylton Saturnino Teixeira (fish)	INPA-Biologia/Limnologia
Prof. Célio Murilo Carvalho Valle (mammals)	University of Minas Gerais, Belo Horizonte
Andy Whittaker (birds)	INPA/WWF Minimum Critical Size of Ecosystems
Jansen A.S. Zuamon (fish)	INPA-Biologia/Limnologia

Technicians

Luis de Sales Aquino	INPA-Entomologia
Nildon Atalino Pinto Ataide	INPA-Biologia/Limnologia
Edimar Ferreira Batista	INPA-Botânica
José Elias Binda Brasil	INPA-Entomologia
Manoel de Santa Brigida	Museu Paraense Emílio Goeldi, Belém
Raimundo Ferreira Cardoso	INPA-Ecologia
Edivaldo Pedroso Carvalho	Government of Roraima
Dionísio Fernando Coelho	INPA-Botânica
Luis Fernando Coelho	INPA-Botânica
José Severino Gomes	INPA
Arnaldo Iran Regio Luz	INPA-Produtos Naturais
Manuel Cursino López	INPA
José Freire Marinho	INPA-Biologia
Francisco Wessen Moreira	INPA-Agronomia
Cosme Damião Alves da Mota	INPA-Botânica
José Benayon Bessa de Moura	INPA-Botânica
André Nilson Neves de Oliveira	INPA-Limnologia
Edson Palheta	INPA-Ecologia
José Francisco Trafano Peixoto	Centro de Ciências de Roraima (CECIR)
João de Souza Pena	INPA-Zoologia
Rosemiro de Souza Pereira	Museu Paraense Emílio Goeldi, Belém
Luis Pardo Portugal	INPA-Biologia de Água Doce
José Ferreira Ramos	INPA-Botânica
José Carlos Raposo	INPA-Biologia/Limnologia
José Ribeiro	INPA-Ecologia
João Rocha	INPA-Ecologia
Pedro Rodrigues Filho	Government of Roraima
Bernadette Tomas Souza Santana	Herbário, Museu Integrado, Roraima
José Lima dos Santos	INPA-Botânica
Cláudio Sena	INPA-Ecologia
Anazildo Mateus de Sena	INPA
Roberto Araújo da Silva	INPA-Biologia de Água Doce
Astrogildo Dias da Silva	INPA-Biologia/Limnologia
Edilson Nunes da Silva	INPA-Ecologia
José Olegário da Silva Filho	INPA-Agronomia
Paulo Gomes da Silva	Government of Roraima
Edinaldo Nelson dos Santos Silva	INPA-Limnologia
Raimundo Carvalho de Sousa	Government of Roraima
Ulysses Oliveira Vieira	Departamento Nacional de Produção Mineral, Manaus
João Vidal Josemar	INPA-Entomologia

2. Land Development

Dr Peter Furley	University of Edinburgh
Dr Christopher Barrow	University College, Swansea
Dr Thomas Dargie	University of Sheffield
Michael Eden	University of London
Dr John Hemming	Royal Geographical Society
Dr Philippe Léna	Museu Paraense Emílio Goeldi, Belém
Dr Duncan McGregor	University of London
Dr Luc Mougeot	Núcleo de Altos Estudos Amazónicos, University of Pará, Belém
Andrew Paterson	Food and Agriculture Organisation
Nelson Vieira	Fundação João Pinheiro, Belo Horizonte

3. Forest Regeneration

Dr John Proctor	University of Stirling
Dr João B. S. Ferraz	INPA-Silvicultura
Dr Valerie Kapos	University of Mérida, Venezuela
Dr Robert Marrs	Institute of Terrestrial Ecology
Robert Miller	INPA-Silvicultura
Duncan Scott	University of Stirling
Dr Jill Thompson	University of Stirling
Virgílio Viana	EMBRAPA-Belterra/Harvard University

Technicians

Estevão	INPA-Silvicultura
Valdivino Valdir Malheiro	INPA-Silvicultura
Gildo de Souza Marcolino	INPA-Silvicultura
Valda Maria Nogueira Ribeiro	INPA-Silvicultura
Antônio Carlos Barbosa Rosa	INPA-Silvicultura
Manuel da Silva	Government of Roraima
Girlene A. Torreias	INPA-Silvicultura
Gildo de Viçosa	Government of Roraima

4. Soils and Hydrology

Prof. John Thornes	University of Bristol
Flávio J. Luizão	INPA-Ecologia
Dr Jo Anderson	University of Exeter
Dra Lucille Marilyn Kriege Antony	INPA-Ecologia
Fabiano Biancucci Apolinário	INPA
Maria Carmozina de Araújo	INPA-Roraima
Sulineide Ataide	Federal University of Amazonas
Arnaldo Carneiro Filho	INPA-Solos
Bridget Gregory	University of Bristol
Maria Regina Costa Luizão	INPA-Ecologia
Dr Stephen Nortcliff	University of Reading
Maria Lúcia Pinheiro da Paz	INPA-Solos
Elizabeth Franklin Ribeiro	INPA-Ecologia
Dr Daniel Robison	University of Reading
Dr Sheila Ross	University of Bristol
Technicians	INPA-Ecologia
Elisiana Pinheiro de Oliveira	Serviço Nacional de Levantamento de Solos
José Francisco Souza Neto	

5. Medical Entomology

Dr Victor Py-Daniel	
Dr Derek Charlwood	INPA-Entomologia
Nelson Antunes	Liverpool School of Tropical Medicine
Dr Toby Barrett	INPA-Entomologia
Dr Sixto Cascaron	INPA-Entomologia
Dr Eloy G. Castellón	University of São Paulo
Assad José Darwich	INPA-Entomologia
Júlio Dellome Filho	INPA-Limnologia
Sandra Tonetti Magni	University of Amazonas, Manaus
Dr José Mauro Martini	Fundação Zoológica do Rio Grande do Sul
Renato Pigoraro	Departamento Nacional de Produção Mineral, Manaus
Ílea Brandão Rodrigues	University of Santa Catarina
Dr Wanderli Pedro Tadei	INPA-Ciências Médicas
	INPA-Entomologia
Technicians	
Ulysses Carvalho Barbosa	INPA-Entomologia
Nelson Ferreira Fé	INPA-Entomologia
Joselias Lima Rodrigues	INPA-Ciências Médicas
Francisco Raimundo Nonato dos Santos	3° Grau, Federal Government of Roraima

João Bosco Magalhães Ferreira dos Santos	INPA-Ciências Médicas
Roberto Stiger	INPA

6. *Administration*

Dr John Hemming (leader)	Royal Geographical Society
Steven Bowles (field leader)	Royal Geographical Society
José Antônio Alves Gomes	INPA-Roraima
Daphne Hanbury (home agent)	Royal Geographical Society
Sarah Latham (nurse)	Royal Geographical Society
Debbie Macklin	ICI-Brasil
Guttemberg Moreno de Oliveira	SEMA-Roraima
Eronias dos Santos Oliveira	SEMA-Roraima
Carlos Paredes	INPA-Roraima
Luis Berto Pestana	
Lorival Araújo dos Santos	INPA-Ecologia
Antônio Alves da Silva	SEMA-Roraima
Patrícia Silva dos Santos	SEMA-Roraima
Amazonas Valquimar Felix de Souza	SEMA-Roraima
Francinete Valquimar Felix de Souza	SEMA-Roraima
Fiona Watson (administrator)	Royal Geographical Society

7. *Steering Committee*

Viscount Montgomery of Alamein, CBE	Chairman
Sir William Harding, KCMG	Lloyds Bank
Tom Long	BAT Industries
Sir John Hall, Bt	Eurobraz Bank
Fred Packard	Banco Garantia
Peter Bareau	Lloyds Bank
Dr John Hemming	Royal Geographical Society
Professor John Thornes	University of Bristol
Dr James Ratter	Royal Botanic Garden, Edinburgh
Dr John Proctor	University of Stirling
Dr Peter Furley	University of Edinburgh

APPENDIX TWO:
SPONSORSHIP OF THE MARACÁ RAINFOREST PROJECT

Grant-giving Organisations & Charitable Trusts

The Ford Foundation
The Headley Trust
Overseas Development Administration
The Baring Foundation
The Royal Society
The Dulverton Trust
Sir Samuel Scott of Yews Trust
The British Academy
Royal Botanic Garden, Edinburgh
Royal Botanic Gardens, Kew
The Royal Geographical Society
Carnegie Trust
The Rank Foundation
The Timber Trade Federation
The Yapp Education and Research Trust
The John S. Cohen Foundation
The Peter Nathan Charitable Trust
The Sir Jules Thorne Charitable Trust
The Frederick Soddy Trust
Lloyds Charities Trust
Anglo-Brazilian Society
The Royal Scottish Geographical Society
The Sir Jeremiah Coleman Trust
The Twenty-Seven Foundation
The Godinton Charitable Trust
The Percy Sladen Memorial Trust
The Schroder Charity Trust
The Leopold de Rothschild Charitable Trust
The Inchcape Charitable Trust
Jocelyn and Katharine Morton Charitable Trust
Sir Andrew Carnworth's Charitable Trust
D M Charitable Trust

Business and Industry

1. CASH DONATIONS:

ICI Brasil
Souza Cruz, Brazil (BAT)
Lloyds Bank International
European-Brazilian Bank, Eurobraz
Linhas-Correntes
English China Clays
Companhia Wilson Sons, Brazil
Gessy-Lever, Brazil (Unilever)
Castrol do Brasil
Peat, Marwick, Mitchell
De La Rue, Brazil
Ciba-Geigy
Grindlays Bank
IMI
Canvey Supply Co.

2. DONATIONS IN KIND: EQUIPMENT

ICI Brasil (boat, refrigerator, medical evacuations)
Mercedes-Benz, Brazil (loan of 13-ton truck)
Philip Harris Holdings (laboratory equipment)
South-Western Marine Factors (Mercury outboard motors)
Booth Steamship Co. (freight)
Sachs-Dolmar (UK) (chainsaws)
Mineração Vale do São João (loan of pick-up)
Petrobrás (petrol and diesel fuel)
Lloyds Bank International (customs guarantee)
Colour Processing Laboratories (film developing)
Souza Cruz (motor bicycle)
British Airways (flight concessions)
Sedgwick International Ltd (insurance concession)

3. FOOD AND OTHER SUPPLIES

Frigorífico Anglo, Brazil (tinned meats)
Banco Garantia and Lojas Americanas (concessions on supplies)
A.P. Burt & Sons
Agfa Gevaert
Baird & Tatlock
Bowater Containers
Bulldog Tools
Colman's of Norwich
Cusson (International)
Davidson Packaging
Dixons (UK)
Dudley Stationery
Envopak (International)
Grangewood Plastic Packaging
Honey Bros
Hawkins & Manwaring
J.B.S. Associates
Lyons Tetley
Marconi Communications Systems
Premier Brands (UK)
Psion
Rotunda
Supreme Plastics
W.H.M. & Partners
William Jones Packaging
Yardley of London

4. MEDICAL SUPPLIES

Abbott Laboratories
Astra Pharmaceuticals
Allen and Hanbury's
Bayer (UK)
Beecham Research
Bencard
Boehringer Ingelheim
Care Laboratories
Ciba-Geigy Pharmaceuticals
Cidal Products
Duncan, Flockhart & Co.
Du Pont (UK)
E.R. Squibb & Sons
Ethicon
Evans
Fisons
Gilrazer
Glaxo Laboratories
Hoechst (UK)
ICI Pharmaceuticals
Immuno
Institut Mérieux (UK)
Janssen
Johnson & Johnson
Kirby-Warwick Pharmaceuticals
Lederle Laboratories
LRC Products
May and Baker Pharmaceuticals
McCarthy Laboratories
Napp Laboratories
Parke-Davis
Pfizer
Phoenix Pharmaceuticals
Rapid Diagnostics
Reckitt and Coleman
Roche Products
Sabre International
Searle Pharmaceuticals
Schering Health Care
Spodefell
Surgikos
T.J. Smith & Nephew
Warner-Lambert (UK)
The Wellcome Foundation
Winthrop Laboratories
Wyeth Laboratories

Individual Contributions

Dr John Bockstoce
Mr Laurence Strenger
The Viscount Montgomery of Alamein
Mr Fred Packard
Dr D.G. Jamison
Brig. Lewis J. Harris
Mr Thomas Pritzker
Mrs Alice Hemming
Dr Donald Bain
Dr John Garver
Mr Tony Watson

REFERENCES

1: *Early Exploration of Maracá*

Report of 1772 by Friar Benito de la Garriga, in Cesáreo de Armellada, *Por la Venezuela indígena de ayer y de hoy* (Caracas, 1960).

Commander Dyonísio Cerqueira, *Relatório sobre os rios Uraricoera e Uraricapara* (1883) in Nabuco, *Limites, Annexes du seconde memoire du Brésil*, 2.

Alexandre Rodrigues Ferreira, *Tratado histórico do Rio Branco* (1786), also in Nabuco, *Limites.*

Report by Captain Ricardo Franco de Almeida Serra and Dr Antonio Pires da Silva Pontes (1781), in Joaquim Nabuco, *Limites entre le Brésil et la Guyane Anglaise* (Rio de Janeiro, 1903).

Manoel da Gama Lobo d'Almada, *Descipção relativa ao Rio Branco e seu territorio* (1787), in Nabuco, *Limites, Annexes du premier mémoire du Brésil*, 1.

Alexander Hamilton-Rice, 'The Rio Branco, Uraricuera, and Parima', *Geographical Journal* (London), 1928, 71:2 and 3.

Theodor Koch-Grünberg, *Vom Roroima zum Orinoco. Ergebnisse einer Reise in Nordbrasilien und Venezuela in den Jahren 1911–1913* (5 vols, Berlin and Stuttgart, 1916–28), 1.

Robert Schomburgk, 'Journey from Fort San Joaquim on the Rio Branco to Roraima, and thence by the rivers Parima and Merewari to Esmeralda on the Orinoco, in 1838–9', *Journal of the Royal Geographical Society* 10, 1840.

The early history of this northern part of Brazil is described in detail in: John Hemming, 'How Brazil acquired Roraima', *Hispanic American Historical Review* (Gainesville, Florida, 1990) 70:2, and *Roraima: Brazil's Northernmost Frontier* (University of London, Institute of Latin American Studies, 1990).

3: *The Land Environment*

Bates, H. W. *The Naturalist on the River Amazons*, John Murray, London, 1873

Kricher, J. C. *A Neotropical Companion: An Introduction to the Animals, Plants and Ecosystems of the New World Tropics*, Princeton University Press, Princeton, New Jersey, 1989

Pires, J. M. and Prance, G. T. 'The Vegetation Types of the Brazilian Amazon' from Prance, G. T. and Lovejoy, T. E., *Key Environments: Amazonia*, Pergamon Press, Oxford, 1985.

Spruce, R. *Notes of a Botanist on the Amazon & Andes*, 2 vols, Macmillan, London, 1908

4: The Aquatic Realm

Fawcett, P. H. *Exploration Fawcett*, Hutchinson, London, 1953
Fleming, P. *Brazilian Adventure*, Jonathan Cape, London, 1933
Gardner, G. *Travels in the Interior of Brazil*, Reeve, Banham & Reeve, London, 1849
Goulding, M. *Amazon: The Flooded Forest*, BBC Books, London, 1989
The Fishes and the Forest: explorations in Amazonian natural history, University of California Press, Berkeley, 1980
Wallace, A. R. *A Narrative of Travels on the Amazon and Rio Negro*, 3rd ed., Ward, Lock, London, 1890

5: Results

The description of Lewis and Milliken finding *Alexa* trees is from Gwilym Lewis, 'Quest for useful legumes' in Nigel Hepper, *Plant Hunting for Kew* (Her Majesty's Stationery Office, London, 1989).